BEE-KEEPING

―FOR―

BEGINNERS.

A PRACTICAL

―AND―

Condensed Treatise ON THE Honey-Bee.

Giving the Best Modes of Management in Order to Secure the Most Profit.

―BY―

J. P. H. BROWN.

Coyrighted.

Copyright © 2013 Read Books Ltd.
This book is copyright and may not be
reproduced or copied in any way without
the express permission of the publisher in writing

British Library Cataloguing-in-Publication Data
A catalogue record for this book is available from the
British Library

Bee Keeping

Beekeeping (or apiculture, from Latin: *apis* 'bee') is quite simply, the maintenance of honey bee colonies. A beekeeper (or apiarist) keeps bees in order to collect their honey and other products that the hive produces (including beeswax, propolis, pollen, and royal jelly), to pollinate crops, or to produce bees for sale to other beekeepers. A location where bees are kept is called an apiary or 'bee yard.' Depictions of humans collecting honey from wild bees date to 15,000 years ago, and efforts to domesticate them are shown in Egyptian art around 4,500 years ago. Simple hives and smoke were used and honey was stored in jars, some of which were found in the tombs of pharaohs such as Tutankhamun.

The beginnings of 'bee domestication' are uncertain, however early evidence points to the use of hives made of hollow logs, wooden boxes, pottery vessels and woven straw baskets. On the walls of the sun temple of Nyuserre Ini (an ancient Egyptian Pharo) from the Fifth Dynasty, 2422 BCE, workers are depicted blowing smoke into hives as they are removing honeycombs. Inscriptions detailing the production of honey have also been found on the tomb of Pabasa (an Egyptian nobleman) from the Twenty-sixth Dynasty (c. 650 BCE), depicting pouring honey in jars and cylindrical hives. Amazingly though, archaeological finds relating to beekeeping have been discovered at Rehov, a Bronze and Iron Age archaeological site in the Jordan Valley, Israel.

Thirty intact hives, made of straw and unbaked clay, were discovered in the ruins of the city, dating from about 900 BCE. The hives were found in orderly rows, three high, in a manner that could have accommodated around 100 hives, held more than 1 million bees and had a potential annual yield of 500 kilograms of honey and 70 kilograms of beeswax!

It wasn't until the eighteenth century that European understanding of the colonies and biology of bees allowed the construction of the moveable comb hive so that honey could be harvested without destroying the entire colony. In this 'Enlightenment' period, natural philosophers undertook the scientific study of bee colonies and began to understand the complex and hidden world of bee biology. Preeminent among these scientific pioneers were Swammerdam, René Antoine Ferchault de Réaumur, Charles Bonnet and the Swiss scientist Francois Huber. Huber was the most prolific however, regarded as 'the father of modern bee science', and was the first man to prove by observation and experiment that queens are physically inseminated by drones outside the confines of hives, usually a great distance away. Huber built improved glass-walled observation hives and sectional hives that could be opened like the leaves of a book. This allowed inspecting individual wax combs and greatly improved direct observation of hive activity. Although he went blind before he was twenty, Huber employed a secretary, Francois Burnens, to make daily observations, conduct careful experiments, and keep accurate notes for more than twenty years.

Early forms of honey collecting entailed the destruction of the entire colony when the honey was harvested. The wild hive was crudely broken into, using smoke to suppress the bees, the honeycombs were torn out and smashed up — along with the eggs, larvae and honey they contained. The liquid honey from the destroyed brood nest was strained through a sieve or basket. This was destructive and unhygienic, but for hunter-gatherer societies this did not matter, since the honey was generally consumed immediately and there were always more wild colonies to exploit. It took until the nineteenth century to revolutionise this aspect of beekeeping practice – when the American, Lorenzo Lorraine Langstroth made practical use of Huber's earlier discovery that there was a specific spatial measurement between the wax combs, later called *the bee space*, which bees do not block with wax, but keep as a free passage. Having determined this bee space (between 5 and 8 mm, or 1/4 to 3/8"), Langstroth then designed a series of wooden frames within a rectangular hive box, carefully maintaining the correct space between successive frames, and found that the bees would build parallel honeycombs in the box without bonding them to each other or to the hive walls.

Modern day beekeeping has remained relatively unchanged. In terms of keeping practice, the first line of protection and care – is always sound knowledge. Beekeepers are usually well versed in the relevant information; biology, behaviour, nutrition - and also wear protective clothing. Novice beekeepers commonly wear gloves and a hooded suit or hat and veil, but some experienced beekeepers elect not to use gloves because they inhibit delicate manipulations. The face and neck are the most important areas to protect (as a sting here will lead to much more pain and swelling than a sting elsewhere), so most beekeepers wear at least a veil. As an interesting note, protective clothing is generally white, and of a smooth material. This is because it provides the maximum differentiation from the colony's natural predators (bears, skunks, etc.), which tend to be dark-coloured and furry. Most beekeepers also use a 'smoker'—a device designed to generate smoke from the incomplete combustion of various fuels. Smoke calms bees; it initiates a feeding response in anticipation of possible hive abandonment due to fire. Smoke also masks alarm pheromones released by guard bees or when bees are squashed in an inspection. The ensuing confusion creates an opportunity for the beekeeper to open the hive and work without triggering a defensive reaction.

Such practices are generally associated with rural locations, and traditional farming endeavours. However, more recently, urban beekeeping has emerged; an attempt to revert to a less industrialized way of obtaining honey by utilizing small-scale colonies that pollinate urban gardens. Urban apiculture has undergone a renaissance in the first decade of the twenty-first century, and urban beekeeping is seen by many as a growing trend, it has recently been legalized in cities where it was previously banned. Paris, Berlin, London, Tokyo, Melbourne and Washington DC are among beekeeping cities. Some have found that 'city bees' are actually healthier than 'rural bees' because there are fewer pesticides and greater biodiversity. Urban bees may fail to find forage, however, and homeowners can use their landscapes to help feed local bee populations by planting flowers that provide nectar and pollen. As is evident from this short introduction, 'Bee-Keeping' is an incredibly ancient practice. We hope the current reader is inspired by this book to be more 'bee aware', whether that's via planting appropriate flowers, keeping bees or merely appreciating! Enjoy.

INTRODUCTION.

I have written "BEE-KEEPING FOR BEGINNERS" not only to supply a want long felt by the bee-keepers of the South, but to promote an industry that adds to the wealth of the country and administers to the comfort and pleasures of its people.

In its preparation I have endeavored to concentrate the gist of the science of modern apiculture, and to embody in it the practical experience of thirty years as a practical bee-keeper. While its modes of practice and methods of manipulation are fully tested for the Southern bee-keeper, its principles can be applied and used wherever the honey-bee is cultivated.

J. P. H. BROWN,
AUGUSTA, GA.

CONTENTS.

CHAPTER I. PAGE

History of Bee-Keeping—Profits and Pleasures—Luck and Pluck—Requisites for Success—Bee Literature 1

CHAPTER II.

Varieties of Bees in the Hive—Workers, Drones and Queen—Undeveloped Females—Duties of Workers—Size of Worker Cells—Length of Life of Workers—Object of the Drones—Length of Life of Drones—Size of Drone Cells—The Queen or Mother Bee—Her Appearance, Sting, Size of Ovaries, and Difference in Size of Abdomen at Different Seasons, and when Placed in Different Size Colonies—Duties of Queen—Fertilized Eggs, Unfertilized Eggs, Number of Eggs a Queen Lays, and Length of Time to Hatch—Larvæ, Royal Food, and Development of Queen—Impregnation of Queens—Parthenogenesis—Spermatheca—Fertile Workers............. 4

CHAPTER III.

Locating an Apiary—The Best Pasture—Distance Bees Can Go for Forage—Arrangement of Hives—Shade—How to Start in the Business—Mistakes of Beginners in Making Purchases—Bee-Keeping for Invalids—Handling Bees—Protection Against Stings—Structure of Sting—Antidotes—How to Open Hives—A Good Smoker Indispensable—How to Use It—Disposition of Different Varieties of Bees—Influence of Color on Temper of Bees—Length of Tongue—Best Honey Gatherers... 13

CHAPTER IV.

Hives—Patents—Langstroth Hive—Size—Top Bars of Frames—Self-Spacing—Narrow Top Bars—Division Boards—Eight Frame Hives—Rabbetted Corners versus Dove-Tail—The People's Hive—Excelsior Langstroth Hive—Paint—Leaky Tops—Loose Bottom Boards................................. 22

CHAPTER V.

Swarming Instinct—Condition of Colony at Time of Swarming—Rearing Drones—Queen Cells—When First Swarms Issue—Hanging Out Not Always an Indication of Swarming—Appearances that Indicate the Time of Swarming—When the Queen Comes out—Clustering—How to Prepare the Hive for the Swarm—Hiving—Straight Combs—How to Separate and Hive Swarms that Cluster Together—Balling Queens—Piping of Young Queens—After Swarms—How to Take Swarms Clustered in High Places—Absconding Swarms—How to Make Swarms Settle—Abnormal Swarms, and Cause of Their Swarming—Swarm Catchers—Ringing Bells....... 27

CHAPTER VI.

Weak Colonies—How to Dispose of Them—Uniting—When to Unite—Feeding and When to Feed—Cost of Feed for Winter Stores—How to Feed—Feeders—How to Feed a Colony in a Starving Condition—Stimulative Feeding—Egg Production and amount of Brood Influenced by the Honey Flow—Advantages of Strong Colonies Rich in Stores in Spring—Individuality of Bees—How to Make Sugar Syrup—Care Required in Feeding—Robbing—What Colonies Do the Robbing—How to Detect Robber Bees—How to Prevent—How to Arrest the Habit When Formed.. 36

CHAPTER VII.

Transferring from Box Hives Into Frame Hives—At What Time to Do It, and Where to Do It—Tools and Implements Required—Transfer Sticks—How to Proceed—The Kind of Comb to Select—How to Get the Bees Into the New Hive—When to Remove the Hive to Its Stand—A Good Plan for Beginners... 42

CHAPTER VIII.

What is Honey?—Honey From Other Sources than Bloom—Adulterations—Do Bees Injure Fruit—Pollen and Its Uses—Propolis—Wax and How Formed—Comb Foundation and How Made—Brood Foundation, How Best to Use It and How to Fasten It in the Frames—Section Foundation, and How to Fasten It... 48

CHAPTER IX.

Which is the Most Profitable, to Run An Apiary for Comb Honey or Extracted? Size of Brood Chamber when Working for Comb Honey—Crate to Hold the Sections—Bee-Space—When to put on Sections—How to Work to the Best Advantage—Wide frames—What Colonies will Work in Sections—How to Dispose of Partially Filled sections—Swarming and Great Surplus Incompatible—Place to Store Sections that are Filled—How to Keep Out the Worms—Fumigation—Best Hive for Extracted Honey—When to Extract—How to Take the Frames of Honey from the Hive—How to Extract—Rules to be Observed—Uncapped Brood Injured—Time When the Extractor Should be Used with Caution 55

CHAPTER X.

Artificial Swarming—How to Make Swarms by Division—Cyprian and Syrian Bees Great Queen Cell Builders—Honey Production and Queen Breeding Antagonistic to Each Other in Practice—Which is the Most Profitable?—The Queen the Prime Factor in the Colony—Capable of Improvement—Highest Type of a Queen—Necessity for Select Breeding Queens and Drones—The Proper Condition of a Colony to Make Good Queen Cells—How to Procure the Eggs to Get the Larvæ—The Right Stage for Use—How to Prepare it and Fix it in the Frame—How to Prepare the Hive—How to Get Bees of the Right Age to make the Queen Cells—How to Keep the Dates—When to Remove the Cells—How to Make Nuclei for the Reception of Cells—How to Insert the Cells—Bees Cutting Cells—How to Prevent—Introducing Virgin Queens—How to Introduce Fertile Queens—Mailing Cages—Candy for Food—How Prepared............................... 64

CHAPTER XI.

Diseases of Bees—Dysentery—Cause—Foul Brood—Appearance—Cause—Treatment—Infection—Treatment of Infected Combs and Hives—Bee Paralysis—Causes—Remedies........ 79

CHAPTER XII.

Enemies of Bees—The Wax Moth—When Introduced into this Country—Description—Its Eggs and Larvae—Galleries and Cocoons—Moth Proof Hives—Mallophora—Braula Coeca or Bee House—Ants and Termites—How to Exterminate—Protection Against Mice—Toads Depredators—Spiders—Birds.. 82

CHAPTER XIII.

Bee Pasturage—Diversity of Mellifluent Plants in the Southern States—How to Form an Estimate of the Honey-Value of a Plant—The Proper Conditions for Honey Secretion—Southern Honey-Flora—Classed as to Value—Honey Resources of Florida—Honey Dew and its Formation...... 88

CHAPTER XIV.

Marketing Honey—The People must be Educated to a full Appreciation of the Uses of Honey—Strained and Extracted Honey—Granulation no sign of Impurity—How to Prepare it for Market—How to Offer it—To Whom to Ship—Glutting the Market............ 96

CHAPTER XV.

Uses of Honey in Medicinal Preparations, in Cooking and in the Arts—Remedies for Diseases of the Mouth, Throat, Bronchi and Lungs—Lagrippe and Colds—Receipts for Honey Cakes, Ginger Snaps, Cookies, Puddings, Vinegar, Metheglin, Mead, &c............ 99

CHAPTER XVI.

Apiary Work Planned for the Year............ 105

Bee=Keeping for Beginners.

CHAPTER I.

HISTORY OF BEE-KEEPING—PROFITS AND PLEASURES—"LUCK"—PLUCK—REQUISITES FOR SUCCESS—BEE LITERATURE.

FROM THE FACT that the hive bee, *apis mellifica*, has been a subject of deep study by the learned in every age, and that apiculture has been successfully conducted by the ancients, it may sound strange when I assert that it has only been within the last fifty years that bee-culture has been developed into a science. Before Huber conducted his observations in the hive, through his assistant, Burnens, the natural history of this wonderful little insect was very imperfectly understood; but since the invention of movable comb hives, the introduction of the Italian bee, honey extractors, comb-foundation, and numerous other appliances to make easy and to facilitate apiarian operations and observations, the economy of the hive is now well understood.

Bee-keeping, when intelligently pursued, affords more profit in proportion to the capital invested, more pleasure, and a greater field for mental exercise than any other of the small rural industries. Poultry keeping soon becomes monotonous. It requires no great amount of judgment to mate up your fowls, to gath the eggs, to set hens, to feed, &c. It soon centres down into a sort of routine task work; but not so with the culture of the honey bee. Here every day gives rise to a new problem for your solution. It is one of nature's grand novels, where the plot is so well laid that the farther you read the more deeply you become interested in the subject.

Bee-keeping is like any other sort of business that is subject to failure and success It does not follow, neither can it be expected, that every one who takes hold of it is going to make it a success. In this business there is no such thing as that abstract something called "*luck*." Luck in bee-culture is always measured by "*pluck*," and by an observance of all those conditions upon which its successful prosecution depends.

Bees can not gather honey unless there is honey in the flowers; and the honey secretion is dependent upon certain atmospheric conditions for its full development; or, in other words, the season must be favorable for the production of a good honey crop, the same as a full crop of cotton, corn, oats, &c.

While it is true that bees "work for nothing and board themselves," it is also true that in order to secure the best results, it is necessary that the labor of these industrious insects be directed by the intelligence and apistical knowledge of the bee-keeper. No one can keep bees profitably without a thorough knowledge of the economy of the hive. He must know what to do, how to do, and when to do; for everything pertaining to bee culture must be done just at the right time.

Along with the purchase of bees, hives and supplies, the beginner should get books treating of the subject, and should subscribe for a bee journal, and read up, and thoroughly study the question as he progresses. He will in this way combine theory and practice and as he proceeds will become proficient and expert with all details of management and manipulation, otherwise he will find it an up-hill business attended with no profit.

CHAPTER II.

VARIETIES OF BEES IN THE HIVE—WORKERS, DRONES AND QUEEN—UNDEVELOPED FEMALES—DUTIES OF WORKERS—SIZE OF WORKER CELLS—LENGTH OF LIFE OF WORKERS—OBJECT OF THE DRONES—LENGTH OF LIFE OF DRONE—SIZE OF DRONE CELLS—THE QUEEN OR MOTHER BEE—HER APPEARANCE—HER STING—SIZE OF OVARIES—DIFFERENCE IN SIZE OF ABDOMEN AT DIFFERENT SEASONS AND WHEN PLACED IN DIFFERENT SIZED COLONIES—DUTIES OF QUEEN—NUMBER OF EGGS—FERTILIZED EGGS—UNFERTILIZED EGGS—LENGTH OF TIME TO HATCH—LARVÆ—FOOD—DEVELOPMENT OF QUEENS—ROYAL FOOD—IMPREGNATION OF QUEENS—PARTHENOGENESIS—SPERMATHECA—FERTILE WORKERS.

IN every normal colony there are worker bees, a queen or mother bee, and during the swarming season, more or less drones or males.

The workers are undeveloped females with ovaries in an imperfect condition. They are the "bone and sinew" of the hive. They do all the work—gather the stores, clean out the hive, secrete the wax, make the comb, feed the young, stand guard at the entrance, ventilate the hive, and when necessary defend their home at the risk of their lives. During the busy season of honey gathering, the length of life of a worker will not average more than sixty days. Those that pass over the winter live much longer. The cells in which the workers are developed are a trifle over one-fifth of an inch in diameter, or, they will average about 50 to the square inch including both sides of the comb.

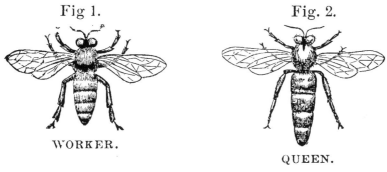

Fig 1. Fig. 2.

WORKER. QUEEN.

Fig. 3.

DRONE.

The drones are large, lusty fellows, rather awkward in their movements, have great power of wing, and and make a loud coarse noise when they fly. Their primary object would seem to be to fertilize the young queens. They have no stings, are great cowards, and are not provided with any organs to gather pollen and honey, therefore nature never intended them to work in the fields. It is the opinion of many observing apiarists, and I think they are correct, that the drones assist in maintaining the normal degree of heat of the colony, and consequently contribute to the curing and evaporation of the honey; but, on the other hand, he is a consumer of honey and not a producer, hence it is advisable for the bee-keeper not to allow too many drones in his colonies. This can be regulated by cutting out the drone comb and replacing with worker comb or foundation.

The drone has a very precarious tenure upon life. In queenless colones or in colonies with old or defective queens, he may live a whole season, while in other cases he may be allowed to live only a few days. When the swarming season is over, normal colonies destroy their drones, and cease to rear others. They also destroy them during a dearth of honey. Hence when you see the workers chasing and pulling the drones out of the hive, you can rest assured that swarming is over with that colony, at least for a time.

The drones are reared in cells larger than those in which the workers are, and when capped over are much more prominent and conspicuous. They will average about thirty two to the square inch counting both sides of the comb. In diamerer they are a trifle more than one-fourth of an inch.

The queen is the mother bee of the colony. In appearance she resembles a wasp more than a worker bee. Her abdomen is a third larger than that of the worker, and when her ovaries are fully developed she presents a graceful and majestic appearance when crawling over the combs. She is provided with a sting, which presents a slight curved appearance, thus indicating that it was intended more for stinging rival queens than other objects. I have handled thousands of queens, but have never been stung by one. The size of the queen's ovaries depends much upon the demand made upon them for eggs. Thus, a queen in a strong colony, other conditions being the same, looks much larger than the same queen would look, after a short time, if placed in a small weak colony where little brood could be reared. For the same reason the abdomen is larger in the spring during the breeding season than it is in the fall when breeding is over.

The universe is made up of wonders. Every object proclaims the touch of an omnipotent hand. The little honey bee bears this divine impress as wonderfully and as powerfully as the largest animal. When we trace the queen through all her stages of development we are most forcibly reminded of this.

It is the mission of the queen to lay the eggs that are to hatch and develop into the perfect inmates of the hive. During the breeding season, when in a large colony, she lays from 2,000 to 3,000 eggs every twenty-four hours. Those eggs that are fertilized or impregnated and deposited in worker cells produce worker bees, and the eggs that are deposited in drone cells *are not* impregnated and hatch out drones. After the queen deposits the egg in the cell she devotes no further attention to it, and leaves it to the care of the workers. In three days, a little sooner if the weather is very warm, it hatches out a tiny little grub or larva. The workers stand ready at this stage to supply it with a whitish gelatinous looking food which they abundantly deposit around it. If a worker larva, it will come forth a perfect worker in twenty-one days counting from laying the egg. It takes twenty-four days for the perfection of a drone. The time is somewhat influenced by the temperature—if very warm, the development is slightly hastened.

When a colony is deprived of its queen, and there are in the hive worker eggs or larvæ not over three days old, the bees go to work to rear another queen from the *just-hatched* larvæ. These little grubs, fed and nursed in the ordinary way, would develop in twenty-one days into common workers ; but the bees intend othewise. They now go to work and enlarge the cell around the little worm-looking mite, and fill up around it with the gelatinous food ; in fact, the embryonic insect literally floats in this substance.

The cell is still enlarged and elongated until about the fifth day from the time the bees started the cell, it is capped over. It now presents the appearance of a ground-pea or goober. The transformation that takes place in this cell is wonderful. Under ordinary feeding, the grub would come forth a worker ; but by the great abundance of the rich food, called royal jelly, deposited in the cell, the anatomical structure of the common worker grub is changed—the abdomen is elongated, the ovaries are completely developed, the sting and mouth organs are changed, the hind legs have not the pollen baskets of the worker, and the very instincts and habits are changed from the worker bee.

Instead of crawling forth a perfect worker in twenty-one days from the egg, she emerges forth a perfect queen in about twelve days from starting the cell, or in sixteen days from the laying of the egg.

The approximate time occupied in the development of the worker, drone and queen bee is given in the following table, tabulated by Frank Benton, in his work on "The Honey Bee":

	EGG.	LARVA.	PUPA.	FROM DEPOSIT OF EGG TO IMAGO.
	Days.	Days.	Days.	Days.
Queen	3	5½	7	15½
Worker	3	5	13	21
Drone	3	6	15	24

In from six to eight days, on an average, the young queen will go out on her "bridal tour," as it is termed, to meet the drone. One trip is usually sufficient, but in case she fails to meet her mate, she will go out again until she succeeds. When fertilized, her ovaries commence to expand and her abdomen increases in size, and in from four to six days she will be laying. If she does not become impregnated inside of seventeen days, she will rarely leave the hive for this object after that, and will be a drone layer. Of course, now and then, there may be an exception to the rule, but all such cases of delayed fertilization are attended with defective ovi-production. Unimpregnated queens can lay eggs, but these unfertilized eggs will only hatch drones. Hence drone eggs are not fertilized with the male sperm. This law of parthenogenesis—of a virgin queen laying eggs that will hatch without impregnation—seems wonderful, but the same law applies to some other species of insects. The fact was first observed and confirmed by Dr. Dzierzon, a celebrated German bee-keeper and scientist.

Fig 4.

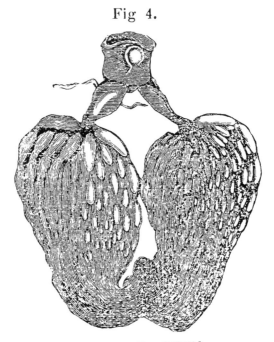

OVARIES OF QUEEN.

After the queen becomes impregnated, it is for life, and she never leaves the hive unless to go out with a swarm. During copulation the male sperm is deposited in a sac called the spermatheca, located on the side of the ovi-duct. The male is killed by the act of copulation. This sac contains sufficient sperm cells or spermatozoa to fertilize all the worker eggs that the queen may lay during her lifetime; and whether the egg laid shall be a fertilized one to produce a worker, or an unfertilized one to produce a drone, is a matter of volition of the queen.

Sometimes in small queenless colonies there are what are termed fertile workers that really do lay eggs that will hatch out drones, and those hatched in worker cells will be of small size. In such cases the bees seem to be so intent on having a queen that some workers by engorgement of food calculated to stimu‑ late the ovaries, become capable of laying eggs. These

Fig. 5.

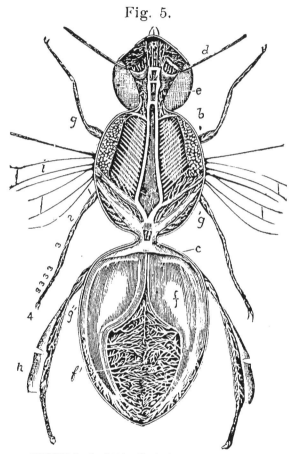

NERVOUS SYSTEM OF THE BEE.

eggs are deposited in the cells without the regularity of those laid by a fertilized queen. Some are placed here and there—lying across each other—piled on top of each other without order or system. In such ab‑ normal colonies there are usually more than one fertile worker. In appearance they can not be distinguished from ordinary workers. The only way to tell them is to catch them in the act of laying. This I have repeatedly done, and have found the underside of the abdomen of these bees a trifle more pendulous than that of the common worker.

CHAPTER III.

LOCATING AN APIARY—THE BEST PASTURE—DISTANCE BEES CAN GO FOR FORAGE—ARRANGEMENT OF HIVES—SHADE—HOW TO START—MISTAKES OF BEGINNERS IN MAKING PURCHASES—BEE-KEEPING FOR INVALIDS—HANDLING BEES—PROTECTION AGAINST STINGS—STRUCTURE OF STING—ANTIDOTES—HOW TO OPEN HIVES—A GOOD SMOKER INDISPENSABLE—HOW TO USE IT—DISPOSITION OF DIFFERENT VARIETIES OF BEES—INFLUENCE OF COLOR ON TEMPER OF BEES—LENGTH OF TONGUE—BEST HONEY GATHERERS.

THERE are very few places in our country where there are no honey yielding plants. The location should, if possible, be near the forage. In our Southern country the best forage is found along the water courses, and in the swamps and bottoms, but on account of malaria that usually abounds in such low places it would be best to locate it on higher ground. One or two miles are not too far for bees to go for forage. I have known them to go four miles, but this distance is too great to enable them to store much surplus.

In our climate hives should be arranged with special reference to shade. I prefer the shade of fruit trees. An arbor of the scuppernong grape vines makes a grand and dense shade. Other varieties of grapes often shed their leaves so soon that they answer a poor purpose. For a shade tree that does not get overly large but has a close compact foliage, I can recommend the *Caradeuc plum*. While it is a shy bearer, the fruit is very delicious. The Mimosa, Chinaberry or Pride of India tree (particularly the low umbrella variety), and *sterculia platanifolia* or varnish tree, are fine for shade, for forage, and very ornamental. These trees are only suited for culture in the Southern States. The honey locust makes a dense shade and at the same time it is a grand honey-yielder. This tree grows well in the Middle States. Where natural shade can not be had, the next best shade is obtained by using pieces of boards for extra covers. The hives should be set with reference to avoid having the hot afternoon sun glaring upon the entrance. In fact it is best not to allow this sun to strike the hive at all; for nearly all the damage to combs by melting is caused by the sun pouring on the hive between the hours of 12 m. and 3 p. m.

When starting to keep bees it is best not to commence with too many colonies. A half dozen in well arranged hives would be sufficient to start with. You must learn to handle them. Practice must be combined with theory. Then, as you gain knowledge you can enlarge your apiary to a profitable size. As knowledge in bee-culture can be conveyed and obtained more rapidly by the eye than by any other means, a few days spent in some well conducted apiary under the direction of an expert apiarist would be of immense benfit.

Beginners often make great mistakes in making their first purchases. Instead of consulting some skilled, practical and reliable bee-keeper as to the articles they need, they resort to some voluminous catalogue of apiarian supplies, and pick out at random articles that are not worth a pewter sixpence to any practical apiarist. Of course their money is spent to no purpose, and quite likely they will soon abandon the business in disgust.

Bee-keeping is often recommended to invalids as a source of outdoor exercise. An invalid might successfully manage a few hives, but it requires a great amount of hard work to properly care for fifty or one hundred colonies. Opening hives, removing frames of comb, hunting queens, extracting honey, forming nuclei &c., &c., is not quite as easy as you might suppose. There is much lifting to be done, stooping over, and often with the hot sun pouring down upon you.

One of the very first requisites towards successful bee-keeping is a knowledge of the nature and temper of the honey-bee, and of the means by which the insect's irascibility can be controlled.

The bee when out foraging never acts on the offensive—always on the defensive. But when its hive is threatened to be disturbed, or its stores taken, or when excited by disagreeable odors, or by persons standing in its range of flight, or by striking at it, or pinching it, &c., its actions may be both defensive and offensive. Its weapon of defense is its sting, which is located at the extremity of its abdomen. This organ does not consist, as many suppose, of one solid pointed body, but is made up of two sections that fit closely together along their edges, while there is a groove running along their inner side through which the poison is injected. The sting is enclosed in a sheath when not in use. Its extremity is barbed, which causes it to stick fast in the object stung. The bee can only release itself by the sting pulling out of its body, along with comes the poison sac and usually some of the intestines. Hence this insect gives also its life along with the blow it strikes.

Fig. 6.

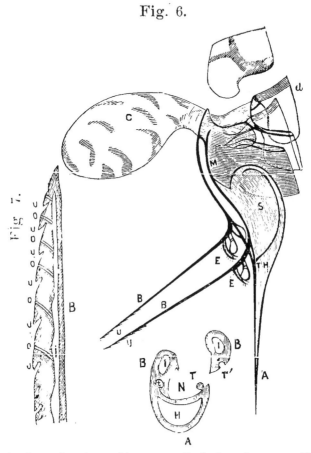

Fig 6. shows the sting and its parts. C. is the poison sac. U. sheath. A. sting proper. B. B. Transverse section of sting. Fig. 7. Barbs of sting very greatly magnified.

When stung, the sting should be immediately removed, or the strong muscles that drive it and still adhere to its base, will force it deeper into the wound.

It should always be scraped out with a dull knife, and not pulled out with the thumb and finger which would squeeze the sac and inject more poison. When I am stung I do not wait for the blade of a knife to remove the sting, but I rub it out with my hand, or if stung on the hand I pass it quickly against a corner of the hive.

In many cases the sting of a bee is attended with much pain and swelling; while in others there are no ill affects produced whatever.

There is no doubt that the system can soon become inured, as it were, to the poison so that no bad effects are produced. The writer well recollects the time when a bee-sting was very painful to him, and was always attended with much pain and swelling; but now he would rather at any time be stung by a bee than be pricked by a thorn.

Many remedies have been suggested for stings, and while they all may do good in some cases, in others they fail. As formic acid is the principle constituent of the poison it seems that those remedies that contain an alkali, such as ammonia, soda, &c., prove effacacious. An application of the tincture of iodine to the wound is said to afford great relief. A continued application of cold water to the part stung is most excellent, and usually prevents the pain and swelling. It can be applied to the part by a wet towel or by pouring from a pitcher.

While it is impossible to work much among bees and never get stung, it is also possible, in most of cases, to keep them "peaceably inclined," so that stings need be "few and far between." One person possesses no more "charms" in handling bees than another, if the same laws and rules are observed.

Nevertheless, the odour or emanations from the bodies of some persons seem to be more disagreeable or exciting to them than those of others.

In all our operations with our bees we must use gentleness. All quick sudden jars and motions irritate them. So does breathing on them. Bees are always more gentle and less inclined to sting when they are gathering honey; and at such times hives can be opened with very little danger; whereas when a dearth of honey prevails the inmates of the same hive might show a great spirit of resentment. I have observed that color exerts some influence on the temper of this insect. I have found that dark shades of clothing, particularly red, are more distasteful to them than white. Hence when working among them I always go in shirt sleeves or in light colored clothes.

When opening a hive always stand on the side opposite to the wind, and never in front of the entrance. It is bee-nature, that when alarmed, for the insect to take to its stores and gorge itself with honey. When in this condition they never sting unless struck at or squeezed. Most writers tell us that this engorgement of the honey sac soothes the anger of the insect and makes it peaceable. While this is partly true, in the main it is incorrect. When the honey receptacle is engorged, the abdomen is much distended, which deprives the bee of making the necessary muscular action of this portion of the body in order to bring the sting into a proper position for a thrust.

Before you start to open a hive it is necessary that you go prepared. I advise a bee-veil for a protection to the face, but it is best to have the hands unprotected, as gloves are much in the way in handling the frames. Rubber gloves are hot and retain the perspiration;

besides, the propolis on the frames soon softens the rubber and destroys them. Woolen gloves covered on the back of hand and fingers with white muslin or homespun answer a pretty good purpose. In order to guard against bees crawling up the arms and legs, it is best to confine the sleeves at the wrist with elastic bands, and to pull the socks over the bottom of pants. It is best not to work among bees after dark, for then they can not see to fly, and "are great at" crawling, and will poke their "noses" in every little opening about the clothing.

A good smoker is indispensable. See that it is in good order. When the fuel in it is well ignited,

Fig. 8.

Direct-Draft Perfect
BINGHAM
BeeSmoker

PATENTED 1878, 1882 and 1892.

approach the hive and blow a few whiffs of smoke in at the entrance. Wait a minute, then blow in a little more smoke until the bees set up a sort of roaring noise. Then gently commence to open the hive, and if the bees show a desire to come up, blow a little more smoke over the tops of the frames, which will run the bees back. Bees can be smoked too much, particularly when queens are to be found. Just how much to give depends upon the humor and disposition of the insect; as a general thing, Hybrids, Syrians and Cyprians require more smoke to subdue them than blacks, Carniolans and Italians. Italians are the most easily handled. When using smoke care should be taken not to run the bees off the combs. Blacks and Syrians are easily run off to the sides of the hive, or will collect in a pendulous mass on the edge or corner of a frame that is being handled, and will possibly loose their hold and drop at your feet, which is not very pleasant to the operator. Cyprians can stand a broadside of smoke unflinchingly, and will only yield after continued blasts. When a bee gets under the clothes, give it room and do not crowd it, and it will make for the light and crawl out without offering to sting.

There are quite a number of varieties of *apis millifica*, among which I may name the black bee, which is the most common. This variety was introduced, it is said, into Pennsylvania from Germany about the year 1627, and was transported to South America in 1845. The Italian, Cyprian, Syrian, Egyptian, Carniolan, &c., are also only varieties, and are undoubtedly of common origin. For beauty, honey gathering capacity, docility, and most desirable qualities, the Italian is to be preferred. In cultivating any of these breeds of bees, there is a continual, though slight disposition to sport from a precise standard of physical and psychical characteristics to an assumption of some of the peculiarities of some other breed. This seems to be a rule attending the breeding of all cattle, horses, sheep, swine, and fancy breeds of poultry, that lack that fixedness and individuality of character sufficient to stamp such breeds as a distinct species.

The Cyprian, Syrian and Egyptian bees are very excitable varieties, and are great fighters and swarmers, and are therefore not desirable.

From a series of experiments that I conducted with delicately adjusted instruments a few years ago with a view to ascertain the length of tongue of the different varieties of the honey bee, I found that the Cyprians and Italians had the longest tongue, and I practically found them the best honey gatherers in my apiary.

CHAPTER IV.

Hives—Patents—Langstroth Hive—Size—Top Bars of Frames—Self-Spacing—Narrow Top Bars—Division Boards—Eight-Frame Hives—Rabbetted Corner Versus Dovetail—The Peoples' Hive—Excelsior Langstroth Hive—Paint—Leeky Tops—Loose Bottoms.

IT is not the province of this little book to go into the history of bee hives, nor to enter into the details of their manufacture, for at this date there is so much competition in the manufacture of bee-keepers' supplies that hives can be purchased either set up or in the flat much cheaper than a bee-keeper could afford to make them himself.

Many years ago the Rev. L. L. Langstroth took out a patent on a bee hive which covered all the valuable features that a hive could possess. This patent is now public property. Since then hundreds of patents have been issued on bee hives, but their claims, in the majority of cases, have been for some little contrivance devoid of value to any practical bee-keeper.

Many modifications have been made of the Langstroth hive. The frame used by Mr. L., was $17\frac{3}{8}$ long by $9\frac{1}{8}$ inches deep; a size that seems to give the most general satisfaction, particularly in the South. In this climate a deeper frame is not desirable for many valid reasons. The top bar of the Langstroth frame is ⅞ inches wide, while the distance between the centre of one comb to the centre of the other is near $1\frac{1}{16}$ inches; this necessitates the spacing of these frames with the fingers. In this way they often get too near together, or too wide apart, particularly at the ends, and the result is crooked comb.

FIG. 9.

EXCELSIOR LANGSTROM HIVE AND FRAME.

In 1870 I adopted a frame with the *ends* of the top bar $1\frac{7}{16}$ inches wide and close fitting. This frame I have used ever since and it gives great satisfaction, as it perfectly spaces the combs. Since then the same style of close-end top bar, with a partial closed end of frame, has been offered to the public as the Hoffman frame. There is another style of Hoffman frame manufactured with the top part of ends close-fitting and a narrow top bar. This style I do not use, nor recommend.

Some bee-keepers contend for a top bar 1⅜ inches wide at ends, but this gives rather too little bee-space between the combs.

With the close end top bar frames it is necessary to have some little space on the side of hive to remove them. To secure this room, some use a division board held by a wedge on the side of the hive. I formerly used this board myself, but it proved very unsatisfactory from the fact that it affords a harbor for spiders, roaches ond other vermin. I find those with a bee-space at the bottom and ends the least objectionable. I nail a narrow top bar to this board and it hangs in the hive as a dummy frame.

A hive holding nine or ten frames is the best. Nine frames may be best suited for most of localities. An eight-frame hive is most too small. If the queen is a good layer she soon crowds the hive and there will be too much swarming, and consequently too little surplus honey.

Fig. 10.

AN EIGHT-FRAME HIVE WITH SELF-SPACING FRAMES.

A nine-frame hive arranged like the People's Hive, with surplus department, and cover proof against leaks, has more advantages and fewer objections than any other. It can be run for either comb or extracted honey; and the section crates will admit of taking sections of different widths.

Fig. 11.

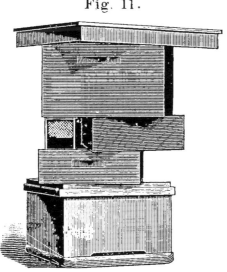

THE PEOPLE'S HIVE.

If the apiary is to be run exclusively for extracted honey, a ten or twelve-frame hive can be used to advantage. The Excelsior Langstroth hive is one of this description. It holds ten frames in the lower story and the same number can be used in the second story, and admits of tiering up one story on top of another.

It is necessary for hives to be well put together in order to stand the continued warm sun of a Southern climate. If not well nailed at the joints they will warp and open. The very best joint is one that is rabbeted and nailed both ways. Such joint is preferable to the locked or dove-tail. Unless kept well painted, wood rots quicker in a warm climate than it does in a cold one.

There is a great diversity of opinion among apiarists in regard to the advantages and disadvantages of fixed or loose bottom boards. The advocates of loose bottoms contend that they are more easily cleaned. This may be correct; but fixed bottoms are the best for a warm climate, and for large apiaries where moving hives is frequently required.

Tops of hives unless covered with tin or some metal will be certain to leak after a few years of exposure to the weather.

As an object for amusement and pleasure, I would recommend an observation or uni-comb hive that contains a single comb with glass on both sides, protected by movable sides or doors. By this arrangement all the movements of the queen and bees can be seen. You can see the queen deposit her eggs, the workers unload themselves of honey and pollen, &c. This hive will hold a small colony that will carry on all the operations of a large one. The hive can be placed on the piazza, or in some shady nook or recess, or in a room, and the bees allowed to pass out and in through a tube in the wall.

CHAPTER V.

Swarming Instinct—Condition of Hives—Rearing Drones—Queen Cells—When First Swarms Issue—Hanging Out Not Always an Indication of Swarming—Appearances That Indicate the Time of Swarming—When the Queen Comes Out—Clustering—How to Prepare the Hive for the Swarm Hiving—When to Put on the Sections—Straight Combs—How to Separate and Hive Swarms That Cluster Together—"Balling" Queens—Piping of Young Queens—After Swarms—How to Take Swarms Clustered in High Places—Clipping Queens Wings—Absconding Swarms—How to Make Swarms Settle—Abnormal Swarms—Cause of Their Swarming—Swarm Catchers—Ringing Bells.

THE instinct to swarm seems to be a desire implanted in the bee to propagate and distribute its race. Wherever the honey-bee is cultivated the swarming season is of intense interest to the bee-keeper. The cry of " *bees swarming !*" with the old time bee-master, was attended with great excitement. It excited the old and the young—the cat and the dog—out come the tin pans the horns and the bells. It was beat, rattle and toot, until the circling, buzzing, roaring insects settled.

In the spring when bees are breeding rapidly and honey is coming in plentifully, the hive becomes very populous, and they make preparations for swarming long before the swarm issues. They rear drones with a view to secure the fertilization of the young queens. Usually after drones are hatched and flying, and drone larvæ capped over, they commence to construct queen cells. After the first queen cells are capped, if the weather is favorable, a swarm may issue. First swarms generally come out in the forenoon, but in case the forenoon should be unfavorable, and they should be ready to issue, they might venture out in the afternoon. In these matters bees do not always follow an invariable rule as many suppose. For instance, there are generally plenty of drones flying before the appearance of a swarm; but I have known swarms to issue when the drone brood was only capped over. Hanging out is not always an indication that the bees are going to swarm. They frequently hang out when they are building comb and rapidly storing honey. Weak and demoralized colonies frequently hang out, and have all the appearance, to a novice, of an over-crowded hive. Too much heat inside may cause them to hang out.

The bee-keeper can not always tell the day the colony is going to swarm by simply looking at the hive; but generally if watched on the morning of the day it is going to cast a swarm, he will find a great many bees flying with their heads toward the entrance, and will pass in and out, not like the workers going to forage, but showing anxiety and a disposition to hurry up the event.

Just before the departure of the swarm the whole colony is greatly excited. The hive seems to be all confusion, bees scampering in all directions over the combs, and filling themselves with honey—now there is a rush for the entrance—old and young—out they come, pell-mell, tumbling over each other, and then circling in the air with a loud roaring noise. The queen may be among the last to get out (particularly if she is an old one); and often comes out, as it were, reluctantly. The workers are the moving spirits in the matter. The idea that the queen leads the swarm is not correct.

After whirling around for some time they will generally settle at some place in a cluster. If the queen is old, or heavily laden with eggs, they usually settle on some low object like a bush, vine, &c.; but if the queen is a young one the swarm may settle at a more elevated place.

In case the queen does not come out with the bees, or they fail to find her, they will return to the hive to come out at another time. When the bees commence to settle you should map out in your mind how you are to proceed in hiving them. Procure your hive and prepare it ready for their reception. Go to the parent hive and take out a frame of brood (see that no queen cells are on it), and place it in the new hive. Take out a frame to make room. This frame can be placed in the old hive in place of the one removed. I would advise filling every other frame in the new hive with sheets of foundation. When arranged thus, it is not so apt to sag and break down by the weight of the bees. Tack a cloth to the edge of the alighting board for a smooth roadway for the bees into the hive. A board, if carefully set up will answer for the cloth.

Before you proceed to hive the bees, sprinkle the cluster well with water, which you can do with a small broom if they are within reach; but if the cluster is high up, put a rose-nozzle on a fountain pump and spray them. This is an instrument that all bee-keepers should possess; it answers an excellent purpose for extinguishing incipient fires, watering plants, &c.

If you can not shake the bees directly in front of the hive, procure a tin bucket and a cloth; and with a whisk broom brush or shake the bees into the bucket; quickly throw the cloth over it and carry it to the hive and tumble them into the hive by first removing the cover, or pour them out in front of the entrance. If the bees are slow to crawl in the hive, take a tuft of grass, a young sprout, or a long handle spoon and stir them toward the entrance. Make them run in lively. Allow no clustering on the outside of the hive—brush them toward the entrance. When the queen goes in the rest will march in like a flock of sheep into the fold. After the great bulk of the swarm is in, you can carry the hive to the place you intend it to stand. This is better than waiting till evening, as is often done. What few bees are out foraging will either find the new location or return to the old hive.

When preparing your hive for the swarm, if you want the bees to confine themselves at first to the brood frames in the lower story, cover the tops of the frames securely with a cloth and over this place the cap or cover. Do not put on the sections or second story, otherwise they might go up into the top of the hive and commence operations. I have known them, owing to a want of care in confining to the lower story, to do this. In two or three days after the bees are hived, they should be looked at to see if they are making their combs straight in the frames. Straight combs are essential in every well-conducted apiary. When you find the bees not making them straight with the comb guide, you must press them in line. It may be necessary to cut some points loose with a knife and secure them with *transfer-sticks*, as done in transferring. At this time place on your sections and give the bees access to them.

Sometimes several swarms may cluster together, and you may want to divide them. In such a case, prepare as many hives as swarms, and place them within easy reach of the cluster. Place an assistant at the entrance of each hive. After sprinkling the cluster, proceed with a long-handle dipper to dip it full of bees from the cluster, and pour in front of one of the hives; then a dipper full in front of another, and so on until you get them equally divided. Your assistants must have each a queen cage at hand and look out for the queens, and see that only one goes into a hive. If more than one is placed before the hive, she must be caged and given to the hive that is queenless. In case several queens should get in the same hive, one will get killed, and sometimes I have known both to get "*balled*" to death by the bees.

Bees have a strange way of getting rid of strange or objectionable queens. They collect around her in an angry mass, forming a ball or cluster, with the poor queen in the center, and bite her wings and pull her legs until they worry her to death. If you wish to release the queen from the angry bees, the safest plan is to drop the ball of bees into a basin of water. When they turn loose, you can pick out the queen. Smoke often makes them more angry. If you use smoke to disperse them, rapidly blow a large volume of it upon the cluster, and as soon as they scatter, pick up the queens. When the queens are valuable and the bee-keeper is unwilling to take any risk, he had better put them in introducing cages, and only release them after the bees settle down to business.

In about six or eight days after the first swarm issues, a second swarm may come out with a young queen recently hatched. Sometimes when several queens are hatching at the same time, they make a peculiar sound called "piping." The noise is like the "peep" of a young chicken, and sometimes can be very distinctly heard if the ear is placed against or near the hive on the evening of the day before the departure of the swarm. One "peep" is a little coarser than the other, and is a reply to that made by a rival queen.

A third and even a fourth swarm frequently come out within a day or two of each other. All swarms after the first are called "casts," and should be discouraged by the bee-keeper, as they often leave the old colony in such a poor condition that it often fails to secure enough honey to winter on. One swarm from a colony is sufficient; best to return such swarms to the parent colony. All swarms after the first have young queens, and as such queens are not very matronly or steady, but good flyers, the swarms may be slow to cluster, and settle in elevated places. In such cases, resort to the small fountain pump; and if the cluster is too high to reach, and the object they are on cannot be removed, you can dislodge them and collect them in a bag prepared as follows: Take a bag or sack and sew an iron hoop (a keg hoop will answer) around the mouth so as to keep it distended. Get a pole sufficiently long to reach the bees; saw a kerf in the upper end deep enough to admit the hoop; crowd your hoop into it. Of course, the hoop will be at right angles with the pole and the mouth of bag open. Now take the pole with the bag, and place the mouth of the bag immediately under the cluster, then with the end of the pole give the limb a good solid lick, which will shake the bees into the bag. Immediately lower the pole, which will also close the mouth of the bag. Shake the bees out in front of the hive. The first time you may not get all the bees; repeat the operation until you secure the bulk of the swarm.

Sometimes a swarm may abscond and take to the woods in spite of all you can do. Casts are more apt to abscond than first swarms. Quite a number of observing bee-keepers contend that bees often send out runners or scouts to secure and prepare a place for the swarm days before it issues In my experience, I have never had any observations that would confirm or reject such conclusions; and I have no right to disbelieve them; but I am pretty certain that the majority of absconding swarms fail to observe these precautions and are not so provident. Swarms that are slow to settle, and show a disposition to abscond may often be brought to cluster by throwing fine dirt in front of them. The reflected light thrown among them from a mirror is said to settle them; also the firing of a gun. These remedies I have never tried.

Clipping the wings of queens is practiced by many bee-keepers in order to prevent swarms from absconding, and to facilitate in operation in the apiary. This clipping is not to be resorted to until after the queen
(3)

is fertilized and becomes an established layer. To perform this operation, the queen should be held by the thorax, between the thumb and fore finger of the left hand, and about one-third or one-half of one large or primary wing on one side cut off with a sharp pointed scissors. Do not hold the queen by the abdomen. There are objections to clipping. Clipped queens, say nothing of the disfigurement, are liable to get lost in the grass when they come out of the hive; besides, when they get on the ground, they are liable to get devoured by ants. Hence it is necessary to be on the look out for such occurrences.

Newly hived swarms frequently refuse to stay in a hive because it is too hot. The empty hive should be kept in the shade, and should be clean and cool when the swarm is put in it, and then it should be protected from the rays of the sun. When the inside of the hives becomes too hot, the bees can not build comb, and must hang out.

Abnormal swarms are those that desert their hives because the bees become demoralized or discouraged for want of stores, have too few bees, &c. Weak colonies in early spring frequently swarm out, and desert their brood, even when they have some stores. Hence in such cases it would seem to be demoralization. You may hive them, but they will swarm out again, probably the same or next day. Your only way to treat such swarms is to unite them with some other weak colonies. Such demoralized bees often get angry at their queen if she refuses to accompany them, and will ball her to death.

To save swarms that may issue when we can not be watching for them, contrivances called "swarm-catchers" and "hivers" have been invented that seek to direct the queen, after she gets out of her hive, into an empty one placed in close proximity to the full one. Were all queens alike in size they could be more easily and more certainly controlled by slots, perforated zinc, &c., but some queens that are very prolific are of small size, and could pass an opening that would admit a worker bee.

The practice of ringing bells, beatings pans, &c., to make swarms settle is a very old but useless custom. It does no good, for they would settle of their own accord. The custom originated in the old country, where there was a law that required you to make a noise in order to notify your neighbors that your bees were swarming, otherwise if the swarm settled on your neighbors' possessions you could not claim it.

CHAPTER VI.

WEAK COLONIES—HOW TO DISPOSE OF THEM—
UNITING—WHEN TO UNITE—FEEDING—WHEN
TO FEED—COST OF FEED FOR WINTER STORES—
HOW TO FEED—FEEDERS—HOW TO FEED A COL-
ONY IN STARVING CONDITION——STIMULATIVE
FEEDING—EGG PRODUCTION AND AMOUNT OF
BROOD INFLUENCED BY HONEY FLOW—ADVAN-
TAGES OF STRONG COLONIES RICH IN STORES
IN SPRING—INDIVIDUALITY OF BEES—HOW TO
MAKE SUGAR SYRUP—CARE REQUIRED IN FEED-
ING—ROBBING—WHAT COLONIES DO THE ROB-
BING—HOW TO DETECT ROBBER BEES—HOW
TO PREVENT—HOW TO ARREST THE HABIT
WHEN FORMED.

IT does not pay to attempt to carry weak colonies over the winter, because after feeding dollars worth of sugar to them they would probably either swarm out or perish before April. Unite as many of these together as required to make a fair colony. If there is any choice in queens, preserve the best. In order to guard against the bees killing her, she had better be caged. Select for use the best of the combs, the nicest and straightest, and the ones that contain the most stores, brood, &c. Smoke both sets of bees until they set up a roaring sound, and then proceed to carefully lift out the frames of the hives with the adhering bees and place them alternately in the hive selected. Shake the bees off the extra combs that you cannot use in front of the hive, and they will enter all right. It is always best to make all such unions late in the evening after sun down ; then they unite peaceably, with no extra flying, and no danger of robbers. To assist them to mark their new home, set a board up in front of the entrance, or place a few weeds up in front until they adapt themselves to the new order of things.

There are many times when it is necessary to feed bees in order to save them from starvation, to build them up, or to promote their breeding. Unless feeding is conducted with judgment, and done at the very time needed, the sugar syrup and the labor are often as good as thrown away. An averaged size colony that is deficient in stores in October will require fully two dollars worth of granulated sugar to carry it over the winter till the first of tne following April, or until the bees can gather plentifully from natural sources.

Fall feeding should be done rapidly, and in as large quantity as the bees can take up at a time. Always place the feed inside the hive. When placed outside, it frequently leads to robbing, a demoralized condition more *easily prevented* than cured. If the nights are warm enough to allow the bees to crawl to the sides of the hive, use a feeder that can be attached to the side or end of the hive and pour in a pint or a quart at night. This feeder is very convenient. It is screwed to the hive on the outside—the feed is poured in from the top—no opening of the hive or disturbing the bees in the least. The bees pass to the feed from the inside of the hive through holes in the hive and corresponding ones in the feeder.

If the weather is too cold for bees to fly out, place

the feed under the quilt over the cluster so they can readily get it. For feeding over the cluster, I prefer an atmospheric feeder with a perforated lid through which the bees suck the feed. When applied to the hive, the lid is turned down over the cluster. The objections are to this feeder; the bees have to be disturbed more or less every time they are fed.

In cases of emergency a feeder may be improvised out of a quart fruit can. Cut off the top, and place in the can a dozen or so of small sticks for floats and supports for the bees to keep them from drowning and falling in the syrup. Nice clean shavings make good floats.

When the colony is in a starving condition, and the bees are only able to crawl, they can best be fed by filling a frame of comb with the feed and placing it gently in the middle of the cluster. The comb can easily be filled by pouring a small stream from the spout of a coffee-pot into the cells. In very early spring we may sometimes find a colony of bees apparently dead, but when the hive is carried into a warm room the bees will gradually revive. Hence never be too fast to condemn bees in this condition.

Many good bee-keepers recommend feeding in the spring to promote breeding. The object is to excite the queen to increase egg-production. This function of the queen is aways governed in a great measure by the honey-flow. When the workers commence to gather honey in the spring the whole hive becomes infused with activity, the eggs in the queen develop rapidly, and she deposits them in a corresponding ratio; but when there is a dearth of honey, egg-production is checked, and there is less brood reared.

As regards the matter of young, there is more wisdom displayed by the honey-bee than by the human. These insects regulate the amount of brood by their ability to sustain and take care of it, but with mankind the number of children is greater among the poor than among the rich.

The object, therefore, of early spring feeding is to stimulate the bees to greater activity, and to deceive the queen into a belief (so to speak) of a honey-flow. When this sort of feeding is started, it must be kept up till the bees can gather plenty of honey from natural sources, for it leads to a great amount of brood which would perish if the feeding be left off without any other supplies at hand.

If your colonies are strong and go into winter quarters, as they should, with from 25 to 30 pounds of honey, this stimulative feeding is not necessary. All you have to do is to uncap the honey in an outside frame, and they will remove and deposit the honey at such points where they most need it. It must be observed that bees have an individuality of their own, and they generally know best how and where to deposit their stores, how much brood they can take care of, when they want to rear drones, etc. It is left for man to assist and guide their instincts to administer to his benefit.

For feeding I use a syrup made by adding one gallon of water to fifteen pounds of granulated sugar. I have never known syrup made from this formula to granulate in the cells. All danger from granulation can be prevented by adding to it two or three pounds of honey. For spring feeding it can be made a little thinner than for use in winter. Always use granulated sugar if you can get it, for most all the brown sugars are adulterated with glucose, a vile preparation which makes it objectional for winter feed. In fact, 40 to 50 per cent. of glucose can be added to brown sugar (and the light shades), and the ordinary consumer cannot detect it. You can use honey for feed by adding one part of water and thoroughly mixing.

When feeding, always use great care not to spill any feed about the hive, or allow vessels containing it to stand near, for all sweets attract the bees, and may lead to robbing and to a demoralization of the whole apiary. Thieving hives of bees are as hard to control as human thieves; and when this habit is once contracted, it causes the bee-keeper an immense amount of work to control it. It is not always the poorer colony robbing the richer one, but more frequently it is the strong colonies robbing the wesker ones. Robber bees can always be told from those that have been out honestly foraging by the peculiar manner in which they approach a hive. Like sneak thieves, they go cautiously with their heads toward the hive, looking for a hole to enter. They will alight at the entrance, and then dart back as if afraid to enter, particularly if there are guards stationed there. But if the entrance is not securely guarded, they will finally pass in, and when once loaded with stolen honey, they will pass out and make for their own hive. The bees belonging to the hive would *come in* loaded and not *go out* loaded. The colony attacked at first may show some resistance, but as the number of robber bees increase, they give up, and frequently will join the robber force to the destruction of the colony. You cannot well arrest the evil until you know the hive from which the robbers come. To be certain, sprinkle some flour on the bees *passing out* of the attacked hive, and have assistants to watch the entrance of the other hives, and the white-coated thieves can be seen entering their hives. When you have found them, smoke them thoroughly in order to alarm them, and to check for the time being their outside operations, and to impart to them the odor of smoke, which will be distasteful to the inmates of the attacked hive. Contract the entrance to the robbed hive so that only one bee can pass at a time ; and set up weeds, grass or boards in front to obstruct the passage way. Robber bees dislike winding entrance ways. Tap a little now and then on the hive to anger the bees and to get them in fighting trim ; but some times they become so discouraged that they loose all desire to defend their home. They nearly always become thus when the robbers have taken all their stores. The only remedy in such cases is to close up the entrance to the weak hive with wire cloth and carry it into a cool dark room, like a cellar ; feed ; allow it to remain forty-eight hours, and then remove it to a new stand taking the precaution to protect the entrance as previously directed. If the hive is still robbed, you had better unite the bees with the dishonest colony. In case where the robbers would attack in force, I have found it of great advantage to spray them well with a fountain pump.

When bees are gathering honey plentifully from natural sources there is no danger from robbing, but when there is a dearth of honey the prudent bee keeper must use every precaution when opening hives, and must not expose frames of comb containing honey to strange bees. Contract the entrances to all weak colonies and place some obstructions in front. An ounce of prevention in this matter is worth more than a pound of cure.

CHAPTER VII.

TRANSFERRING FROM BOX HIVES INTO FRAME HIVES—AT WHAT TIME TO DO IT—WHERE TO DO IT—TOOLS AND IMPLEMENTS REQUIRED—TRANSFER STICKS—HOW TO PROCEED—THE KIND OF COMB TO SELECT—HOW TO GET THE BEES IN THE NEW HIVE—WHERE THERE ARE MANY COLONIES TO TRANSFER—WHEN TO REMOVE THE NEW HIVE TO ITS STAND—A GOOD PLAN FOR BEGINNERS.

IT is often necessary to transfer bees and combs from old box hives, &c., into frame hives. This is a very simple operation, though laborious. The inexperienced should never perform this operation out of doors except at times when there is a honey-flow; for if done when the bees are not gathering, the transferred colony would incur the risk of being captured by robbers.

In order to secure both increase and surplus honey from the transferred colony, the operation should be performed as early in the spring as practicable. The expert can transfer at any time in the year when the bees can fly, but the tyro had better wait till about the time of the apple bloom.

I prefer to make my transfers in a close room with one window. Where there are more windows they can be shaded. There must be no place for bees to get out, nor place for strange bees to get in. In this room I want a table, a small box without top, a

BEE-KEEPING FOR BEGINNERS. 43

hatchet, cold chisel, saw, a long-bladed knife, a bucket of water, a basin and a towel. In front of the window set your new hive with the frames removed, and the entrance raised an inch higher than the back, so as to prevent any drip of honey from running toward the entrance and smearing it up, which would prevent the bees from going in. Set the table near the middle of the room. Have at hand ready prepared a lot of transfer clamps or sticks to hold the combs in the frames. These sticks should be one half inch longer than the depth of the frame; or, in other words, their ends should project a full fourth of an inch above the top bar and the same distance below the bottom bar. They can easily be split from straight-grained pitch yellow pine. The ends of these sticks are notched to hold small, thin wires that keep the sticks in position on the comb. They are used in pairs—two ends are wired together with a space between the width of the bottom bar; while one of the other ends has a wire three or four inches long to be wrapped around the end of the opposite stick when the pair is adjusted to the frame.

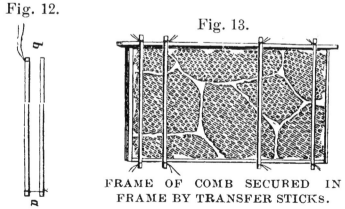

Fig. 12.

Fig. 13.

FRAME OF COMB SECURED IN FRAME BY TRANSFER STICKS.

TRANSFER STICKS.

It generally takes from three to four pair of these sticks to a frame. When all is ready, go to the hive to be transferred, and turn it mouth up; throw a cloth over it and carry it into the room and set it in front of the table, entrance end up. The table should have a cloth on, with one side hanging down, extending a little below the top of the hive. Set the hive close up against the end of cloth. Place the box on the table with the open side down, and allow it to project one-third over the edge of table and directly above the hive to receive the bees that crawl up the cloth. Blow smoke across the top of the hive to keep the bees down, and remove the cloth covering the mouth of the hive and spread it over the box on the table, allowing one edge to hang over the top of the hive. Many persons drum with sticks on the hive to alarm the bees and to get them to ascend into the box. This drumming will do in warm weather, but when the day is cool and the colony not very strong, the bees will not ascend and the time spent in drumming is lost

Blow plenty of smoke under the cloth, across the top of the hive and combs to keep the bees back, and to alarm them. When alarmed, they fill themselves with honey and commence to crawl up toward the top of the hive. Now remove the cloth that you have thrown over the hive and box and proceed with the saw to cut the cross sticks. Then take chisel and hatchet and remove one side of the hive—the side that runs the nearest parallel with the combs. Use smoke, whenever the bees come in your way, to drive them back. When the side is off, take the knife and cut around the edge of the comb and loosen it from the hive, and carefully remove it and lay it on wide boards or on the table. If any bees are on brush them off gently with a whisk broom. Proceed in this way until the combs are all removed. After you get all the combs out, you have to go to work to place them in the frames. First lay your transfer sticks across a board (called a transfer board); then place a piece of comb on top and put your frame down and cut the comb to fit. Press it in the frame, and bring your stick across and wrap the wire. Now hang the frame containing the comb in the new hive. Proceed in this way until all the comb is fixed in the frames. Use only the straight worker comb, and reject the drone and all crooked pieces that will not make a nice fit in the frame. When inserting the combs in the frames, try to preserve the same order the comb occupied in the old hive. If you examine comb you will find the cells incline a trifle upward. Place the brood in the centre and the honey and pollen on the outside. It is best not to place too much comb honey in the frames, for there may be too much drip, and if you get the entrance all smeared and clogged with honey the bees will not enter. Save all the bits of comb containing honey—for the bees may need it—and place it in dishes to be fed in the second story; or, if extracted or squeezed out of the combs, it can be placed in feeders. Wait a couple of days before you feed, until the bees clean up, and fasten the combs in the frames with wax. After the bees wax the combs in the frames, remove all the transfer sticks.

Some bee-keepers use a mixture of two-parts of wax and one-part of rosin to hold the combs in the frames. This works very nicely with dry comb, but will not do with comb containing honey or much brood.

After the frames are all in the hive, securely cover them with the quilt so no bees can get above them into the second story; for sometimes they will crawl up into the cap of the hive and commence to make comb there, and desert the lower story containing the transferred combs. This you must guard against. Now take the clustered bees in the box on the table-cover, and shake them in front of the hive on a cloth that should be tacked to the edge of the alighting board. Stir them in like a swarm. The bees that cluster on the window, brush down into a pan and throw them in front of the hive. When you are transferring a number of colonies it is not necessary for you to wait for all the bees to get in before you can start on another hive. What few are out will go in with the following hive. Always be careful to have the queen in her own colony.

If the weather is quite cool the bees had better be confined with wire-cloth, and carried into a dark room for 24 or 48 hours, until they clean up things, before you set them out. If the weather is warm, and there is a flow of honey, set them out late in the evening of the day you transferred them. Take the precaution to contract the entrance and place some obstructions in front to ward off strange bees. If the colony is short of stores, you must liberally feed; for in restoring and straightening up their combs and in rearing brood they will need large quantities.

The following is a good plan for beginners: Allow one swarm to issue from your box hive. Hive this in your frame hive. Then in about twenty-one days after the issue of the swarm, transfer the old colony. The parent colony will now have a laying queen, whereas had you transferred just after the departure of the swarm, you would run the risk of destroying all the queen cells, and probaby loose the colony. There are other methods of transferring, but in this treatise, I have only given the plan which I consider the most practical and meritorious.

CHAPTER VIII.

WHAT IS HONEY?—HONEY FROM OTHER SOURCES THAN BLOOM—ADULTERATIONS—DO BEES INJURE FRUIT?—POLLEN AND ITS USES—PROPOLIS—WAX—HOW FORMED—COMB FOUNDATION—HOW MADE—BROOD FOUNDATION AND HOW TO FASTEN IT IN THE FRAME—HOW BEST TO USE IT—SECTION FOUNDATION AND HOW TO FASTEN IT.

HONEY is a sweet substance secreted by the nectaries of flowers. It is also secreted in small quantities by little glandular organs on certain plants like the cow-pea. It is taken up by the proboscis of the bee, and deposited in a special pouch called the honey sac, and conveyed to the hive. It is possible that the insect imparts to it some little acid; but honey is not nectar digested by the bees, as some writers contend. The odor, flavor and qualities of it depend upon the source from which it is gathered. Thus, the famous honey of Hymettus has its thyme odor and flavor; the horse-mint honey has its distinguishing qualities; the sage, the poplar, the clover, &c., have their characteristic flavor.

When natural forage is scarce, bees will gather sweets from many sources. They will collect the excretion of the aphis, the waste of cider mills, cane mills, the refuse of molasses and sugar barrels, fruit juice, &c. But none of such stuff can truly be called honey. Bee-keepers have often been charged with feeding their bees glucose, sugar, &c., for the object of the insects storing it in the surplus department to be sold for genuine honey. But such charges cannot be sustained. In fact, for a bee-keeper to purchase sugar, glucose, &c., to feed to his bees with the expectation of their depositing it as surplus honey to be sold at the price of the true article, would be to incur a loss in dollars and cents. This has repeatedly been tried, but only at a pecuniary loss

In some seasons when there is a great scarcity of honey in the flowers bees will work on fruit, but it is only when put to great straits for food that they will attack sound fruit. When the skin of grapes, peaches, figs, &c., become punctured or cut by wasps, yellow jackets, birds, &c., they will proceed to suck the juice. But as all such cracked fruit is unsable they do comparatively little damage. Here it should be remembered that the bee is the friend of the agriculturist— for if it were not for these insects, the fertilizing element of many male flowers would fail to reach the pistils of the female and consequently the plant would produce neither seeds nor fruit. Hence the bee is one of nature's great pollenizers.

Honey is the chief food of the mature bees, but when breeding, they also consume large quantities of pollen which they use in connection with honey in preparing the food for the larvæ.

Pollen is a farinaceous dust adhering to the anthers of flowers, and contains their fertilizing element. Bees collect this in little hairy baskets on their hind legs; and when the little insect gets its load well packed, it has a yellow ball on each hind leg. When working in some trumpet shaped flowers like the yellow jesmine, cotton bloom, &c., it frequently gets its back also coated with pollen presenting the appearance of having been in a meal barrel. Some of the ancient bee-keepers thought that pollen was used to make the wax, but such is not the case. It exerts a very stimulating effect upon all the inmates of the hive. Very little breeding is done before pollen can be gathered in the spring, but as soon as the bees commence to carry it in, the queen's ovaries enlarge, and breeding goes on rapidly. Pollen is the principal food of the young bees; it is digested along with some honey by the workers into a sort of chyle, and in this form, deposited in the cell around the larvæ Rye flour, and other farinaceous substances, have been substituted and fed for natural pollen, but all such experiments have not been attended with any great advantages.

Propolis, often called bee-glue, is gathered by the bees to close up cracks and crevices about the hive. They collect it from various trees yielding gum and rosin. During the honey flow, they pay little attention to this substance, but when the pasturage becomes scant they commence to collect propolis and glue up all the little openings about the hive that would let in wind, rain and insects. They also propolize all inside cracks that will not admit of the passage of their bodies.

Wax is a sort of bee fat that is secreted by organs along the abdomen, and comes out from under the abdominal segments in thin flakes which are removed by attending bees and, after kneading with saliva, are worked into the wonderful structure of comb. Bees, when necessitated, will make comb at any time when the temperature is sufficiently warm to manipulate the wax; but they always work at it with more vim and earnestness about swarming time. When deprived of honey bees cannot make wax. It has been determined by Huber and other experimenters that it takes from ten to fifteen pounds of honey to make one pound of wax. This has been ascertained by taking all the honey and combs from the bees, and then placing them, confined, in an empty hive, and feeding a given weight of honey, then the combs built were also weighed. In all such experiments the results can only be approximately correct, yet they are sufficiently accurate to prove that there would be a great saving of time and honey if the bees were furnished combs already prepared, or comb foundation for the base of the comb.

When working in the apiary it is best to carry along a light box to put all the bits of comb in. Press it into hard balls and then the worms will not so readily destroy it. These pieces of comb, when saved, can be rendered into nice wax which will always command a good price. The handiest instrument for this purpose is a solar wax extractor. But when you have ten or more pounds to render, the quickest way is to use a wash-pot. Fill your pot one-fourth full of water and bring to a boil. Reduce your fire to a slow one. Put comb into a coarse bag, gage the quantity to suit the pot, and place the bag into the pot and weight it down with a stone. Keep up a slow fire. As the wax comes on top dip it off into a vessel of hot water. Keep plenty of water in the pot or the wax may burn. Use pressure on the bag to get out the wax. After you get out all the wax you can, re-melt it in water and pour into cakes. Old rusty iron and tin will discolor wax and spoil it.

Comb foundation is made by passing sheets of pure beeswax through rollers whose surface is impressed or embossed with the base of the cells of the comb. The best brood foundation should run near six square feet to the pound, which, if well fastened into the frame, will not sag.

Fig. 14.

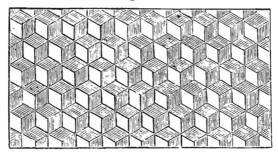

FOUNDATION.

When thinner than this it will sometimes sag unless it is secured in the frame by thin wires running perpendicularly or diagonally across it. These wires are so objectionable to the bees that they spend much time in trying to cut them out. Brood frames that have a triangular comb guide will take foundatian very securely. The best plan that I have found to fasten foundation into these frames is, to get a board that will fit inside of the frame and half as thick as the width of the top bar; lay your frame down and place your board in it, and lay your sheet of foundation on top with three-eights inches of top edge overlapping the comb guide. Now, with your thumb and a thick bladed knife, press and rub the edge of foundation down fast to the wood. It is not necessary to fasten it to the ends of the frames, neither should it touch the bottom bar by a half inch.

It is not best to place swarms in hives with all the frames filled with foundation, for there would be a liability for it to break down by the weight of the bees, particularly if the weather should be very warm. My plan is to alternate the frames of foundation with frames of comb, or if the latter cannot be had, with empty frames. Then the bees will have a support aside from the foundation, which they will soon firmly secure with wax to the frame.

The use of foundation always enables the bee-keeper to secure nice, straight worker combs, which is a consideration of great importance in every apiary; besides it places him in a position to control almost wholly the number of drones which is of great importance in breeding queens.

It is always best to use full sheets in the frames, but narrow strips of an inch or two wide can be used to great advantange. The bees draw out the cells of foundation with astonishing rapidity. I have had them to draw whole frames of it out nearly complete in twenty-four hours, and in that time the queen had the cells filled with eggs.

Thin foundation for the surplus department usually runs from 9 to 10 square feet to the pound Some is made with flat bottom cells which runs from 10 to 12 square feet to the pound, but the bees do not take to this flat bottomed foundation as readily as they do to that which has the natural base of the cell.

Some bee-keepers of note use only starters of this thin foundation in the sections; but it is best to fill the sections one-half or two-thirds full. I find this a gain in time and honey.

Quite a number of contrivances have been invented to fasten the starters of foundation in the sections, but I have adopted the following plan as the most satisfactory to myself, viz: First, cut your foundation into pieces of the size you want; then have at hand a pan of wax over a small kerosene heater, and when melted dip one edge of your foundation in and immediately apply to the section. To do this properly will need some little practice. See that the foundation is well fastened, for if it should get loose the bees would fasten it at the point where it dropped, and would probably make crooked comb in all the adjoining sections.

Parker's machine for fastening starters in sections is probably as efficient and rapid as any other. It is simple and cheap, and can be made to do the work well. To prevent the lever from sticking to the foundation, keep it moistened with honey.

CHAPTER IX.

WHICH IS MORE PROFITABLE, TO RUN AN APIARY FOR COMB HONEY OR EXTRACTED HONEY?—SIZE OF BROOD CHAMBER WHEN WORKING FOR COMB HONEY—CRATE TO HOLD THE SECTIONS—BEE SPACE—WHEN TO PUT ON THE SECTIONS—HOW TO WORK TO THE BEST ADVANTAGE—WIDE FRAMES—WHAT COLONIES WILL WORK IN SECTIONS—HOW TO DISPOSE OF PARTIALLY FILLED SECTIONS—SWARMING AND GREAT SURPLUS INCOMPATIBLE—PLACE TO STORE SECTIONS THAT ARE FILLED—HOW TO KEEP OUT THE WORMS—FUMIGATION—BEST HIVE FOR EXTRACTED HONEY—WHEN TO EXTRACT—HOW TO TAKE THE FRAMES OF HONEY FROM THE HIVE—HOW TO EXTRACT—RULES TO BE OBSERVED—UNCAPPED BROOD INJURED—TIME WHEN THE EXTRACTOR SHOULD BE USED WITH CAUTION.

WHETHER it is more profitable to run an apiary for extracted honey than for comb honey depends much upon the location and the market. Taking one season with another, considerably more extracted honey can be taken than comb honey, but when we get in the market probably one-fourth more for our comb honey than we do for the extracted article, the difference is not much. It is always best to have both kinds to suit all customers.

When we work for comb honey it is best not to have the brood chamber too large, for when there is great space here the bees will not go into the sections as long as they can find room below for the surplus. A hive with nine or ten frames is large enough for any location in the South. The one described in chapter fourth, is the one I recommend for comb honey. You can place on it one or two crates of sections as your pasturage and bees will warrant. The hive should sit level on its stand; for if it leans to one side the foundation will get out of plumb and crooked and bulging comb will be the result.

Fig. 15.

ONE-PIECE SECTION.

Fig. 16.

Section with Foundation Starter.

The bee-space being at the bottom, you can use any width of section. If they do not fill out, place in a dummy. A crate with a bee-space below has great advantages over one that has the space on top; because with the former, you can use any width of section, while the latter will admit of only one width. Besides, with this crate you do not need slatted honey boards nor separators.

Some years ago there was a plain section without a bee passage brought to notice in the bee journals. The bee space was secured by means of a slotted separator. At the time very little attention was paid to the device, and it dropped out of sight. Recently the old idea has been resurrected in new habiliments with a slotted separator called a "fence." It is claimed the honey secured in these plain sections is nicer and better capped than in the standard style of section.

Whether the trifle of gain (if any) in appearance of the honey would make up for the extra cost of these fence separators is a question which is not yet determined.

The sections should be made ready before you want to use them, and as soon as you see the bees adding wax to the upper part of the brood combs, put on your tier of sections. When they commence to gather surplus honey they bulge out the brood combs at the upper part with nice white comb and fill it with honey. If there is a good honey flow it does not take them long to fill the crate. When it is about half complete, raise it up and place an empty one under it next the brood chamber. The bees will not like the space between their surplus and the brood, and will tax their energies the more to fill it. As soon as the first or top tier of sections is complete remove it before the bees discolor the capping of the honey by crawling over it. It is best to remove the sections as soon as capped if we wish to preserve them white and discolored. The middle sections are usually finished before the outside ones; and if you wait till these are capped, the middle ones will be made dark by the bees crawling over them with their pollen-stained feet. Put empty sections in the place of the ones removed. If the honey harvest will justify it, you can raise the second crate and place an empty one under as before. By this plan, carefully managed, I am satisfied you can obtain more surplus comb honey than by any other.

Some very good bee-keepers use wide frames to hold the sections, a plan that I do not recommend. A full depth frame will hold eight sections $4\frac{1}{4}\times4\frac{1}{4}$, a half frame only four sections.

It is necessary to use separators with these frames if you want the bees to keep the combs straight in the sections. You suspend the frames in the upper story. You can get six in the upper story of an eight-frame hive, and seven in the upper story of a ten-frame one. In the midst of the honey flow you can insert one of these wide frames on each side of the brood chamber, of a ten-frame hive, and confine the brood to seven frames.

In the greater portion of the South the bulk of the surplus honey is gathered by first swarms. Mr. Hedden and, I believe, Mr. Hutchinson, both experienced bee-keers of Michigan, contend that in their location it is best to hive swarms that issue in the midst of the honey flow, either on full sheets of foundation or on foundation starters, and not on full frames of comb already worked out. That when the latter is given them they fill the complete cells with honey much faster than the queen can deposit eggs in them; but on the other hand, that when they have to draw out the cells of foundation or construct the comb, the queen deposits the egg in the cell before it is ready for honey, and consequently there is more compulsion for the bees to place the honey in the sections.

A strong colony will generally work in the sections before it swarms. Some sections they may complete and cap over, others remain incomplete; but as soon as the first swarm issues the section boxes are deserted, and if second swarms issue work in that crate of sections is stopped for the season so far as that colony is concerned. In order to get these sections complete it is necessary to practice a *coup d'etat* movement.

As soon as the new swarm is well to work and established in its new hive, which will be in two or three days, remove the crate from the old hive and place it on the new colony, and they will soon complete the job. The new colony has the queen, the vim and the energy, while the old parent hive, for a time, has no laying queen, and is greatly reduced in its working force.

Over-swarming is the bane of the bee-keeper. One swarm is as much as a colony should cast consistent with surplus for the owner. Our *full* honey crop comes only from our strong colonies, and these must be strong *at the time of the honey flow*. To secure this condition, the beginner must get his colonies in good shape in the fall with twenty-five or thirty pounds o: stores, and then, if the queen is vigorous, the colony will come out early in the spring with an army of workers ready for the harvest. Many plans have been suggested for the prevention of swarming, but really I know of no certain one that will meet all cases. When bees get the swarming fever, they will often take it into their heads to swarm in spite of all plans.

Cutting out queen cells, removing frames of brood and giving frames of foundation, often prevent it for the season. This plan works better in the prevention of after-swarms than it does with first swarms. In this case, cut out all queens' cells but one. If "casts" issue, return them to the parent hive after removing a frame or two of comb and substituting foundation. Cage queens when there is danger of their being killed. When working for extracted honey, swarming can usually be controlled or prevented by close extraction and the removal of a frame or two of brood and replace by empty frames.

The filled sections should be stored in a close, tight,

dry room, free from dampness. When comb honey is stored in a damp place, the honey in the uncapped cells will absorb moisture and ferment. In a few weeks' time the combs should be examined to see if any wax worms are in. You can easily tell by the flour-like dust which they leave on the comb. You can only kill them by fumigating the combs of honey with sulphur. To do this successfully the room must be air-tight. Place an iron pot on some brick in the room, and put in a shovel full of live coals, and drop on top of the coals a table spoon full of flour of sulphur, and immediately close the room for twenty-four hours. This amount of sulphur is sufficient for a room ten feet square. One application will usually be sufficient. The honey is not affected by the fumigation. If the room is kept close enough to exclude the wax moth there will be no further danger, as all the eggs have hatched and the larvæ killed. When the amount of comb honey is not large it can be fumigated in a large dry goods box. Nail supports around the sides of the box to hold the sections. Cut a hole in the box near the bottom large enough to admit a small iron pan or skillet, and furnish it with a sliding door. Also place at bottom a few brick to keep your pan off the wood. Place your honey in and cover the box tight; have all cracks securely closed so no air can get in; now place a few live coals in your pan and sprinkle on a teaspoonful of sulphur; quickly pass it through the hole into the box and close the door. Frames and sections with empty comb can be treated in the same way to keep out the worms.

Sections that are only partially filled with honey and not completed had better be extracted, for if the comb is nice and white, they can be used the next season to great advantage.

When we work for extracted honey it is best to have a hive to hold from ten to twenty frames. Some beekeepers extract the honey before it is properly cured in the comb, and before it is capped over, and then after extracting they allow it to stand for some time in shallow vessels or tanks to evaporate. These beekeepers very often have the mortification to find that their honey sometimes ferments and sours, and becomes unsaleable. The *safe plan* is to allow the bees to pretty well cure it in the hive before extracting. Half of the comb should at least be capped over, then the honey will keep in any climate.

Fig. 17.

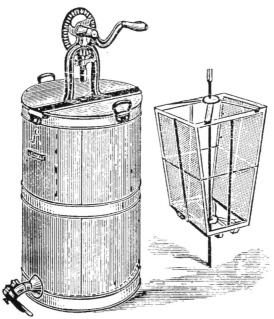

MUTH'S HONEY EXTRACTOR.

The honey extractor is a machine made to sling the

honey out of the cells by centrifugal force. Before the frames of honey are placed in the revolving reel of the extractor the cells must all be uncapped with a knife made for the purpose.

Fig. 18.

MUTH'S UNCAPPING KNIFE.

Fig. 19.

BINGHAM'S UNCAPPING KNIFE.

The cappings should be placed in a vessel like a colander, so the honey can drain from them. The velocity of the machine should only be sufficient to throw the honey from the cells, then there is less danger of breaking the combs. If any comb should break, use transfer sticks to hold it in place till the bees can repair it. The extracting room should be made close with wire-cloth door and window screens so no bees can get in to annoy or rob.

A set of extra combs should be used to place in the hive in place of the frames of honey removed. This saves time and danger from robbers when opening hives. After the first combs are extracted they should be used to replace full ones from another hive.

You should have a light box with a tight lid to hold the frames when removed from the hive and to carry them back and forth. When you remove the bees from the frame of comb, stand at the back of the hive and lean over to the front with the frame, and with a quick motion shake the bees off on the alighting board. In this position there is less danger of getting stung than if you stand in front. Brush off the remaining bees with a tuft of grass or the sprout from a peach tree. A wing or feather irritates them very much and is not as good as the grass.

After the honey is extracted it should stand for a few days in some deep vessel in order to allow the small particles of wax to rise to the top, to be skimmed off before it is placed in permanent vessels.

While the extractor is an indispensable machine to the successful bee-keeper, its use can be abused. It is always well to know when to extract and when to stop; what frames to extract and what to let alone. When the honey flow is very abundant, the same frames can be extracted every five or six days, or as often as the bees get them filled.

Years ago, when I first commenced to use the extractor, I was told, by *the then bee-keeping lights*, that uncapped brood was not injured by it, but I soon found out that it was, and I went more cautiously. Frames containing such brood should not be extracted. When the brood is capped over I do not think it is injured by the operation.

It should be borne in mind that in most of the Southern States the spring honey flow is over by the middle of June, and that July, August and a portion of September are dull months and hardly afford sufficient forage to sustain the bees; hence do not extract too closely, otherwise your bees will be set back and probably may not be able to gather a sufficient support for winter.

CHAPTER X.

ARTIFICIAL SWARMING—HOW TO MAKE SWARMS BY DIVISION—CYPRIAN AND SYRIAN BEES GREAT CELL BUILDERS—HONEY PRODUCTION AND QUEEN BREEDING ANTAGONISTIC TO EACH OTHER IN PRACTICE—WHICH IS THE MOST PROFITABLE?—THE QUEEN THE PRIME FACTOR IN THE COLONY—CAPABLE OF IMPROVEMENT—HIGHEST TYPE OF QUEEN—NECESSITY FOR BREEDING QUEENS AND DRONES—THE PROPER CONDITIONS OF A COLONY TO MAKE GOOD QUEEN CELLS—HOW TO PROCURE THE EGGS TO GET THE LARVÆ—THE RIGHT STAGE FOR USE—HOW TO PREPARE IT AND FIX IT IN THE FRAME FOR USE—HOW TO PREPARE THE HIVE—HOW TO GET BEES OF RIGHT AGE TO MAKE THE QUEEN CELLS—HOW TO KEEP THE DATES—WHEN TO REMOVE THE CELLS—HOW TO MAKE NUCLEI FOR THE RECEPTION OF CELLS—HOW TO INSERT THE CELLS—BEES CUTTING THE CELLS—HOW TO PREVENT—INTRODUCING VIRGIN QUEENS—HOW TO INTRODUCE FERTILE QUEENS—MAILING CAGES—CANDY FOR FOOD—HOW PREPARED.

WHILE the majority of the largest honey producers of our country prefer that their bees should swarm naturally, there are locations and cases where artificial swarming can be practiced to great advantage. There are many plans to do this, and each one, in the hands of the bee-keeper who follows it, *is the best*. Now I will give a plan or two which I think best for the beginner:

The practice of the science of bee-keeping is not unlike the pursuit of other branches of science. You gather an idea, and that idea enlarges and begets other ideas that enable you to generalize your science and modify its modes of application.

I now suppose that your apiary consists of one colony and you want to divide it, and do not care to take the chance of losing the swarm. But you must not think of doing it until it is near the eve of swarming. The hive should be "boiling over" with bees, drones flying, and honey coming in. Get your empty hive, and go to the hive you want to divide and draw from it two frames of brood, and one of pollen, and stores with all the adhering bees, and place them in the new hive. One of theae frames should have the queen on ; for you wish to put her also in the new hive. Shake two-thirds of the remaining bees in the in the old hive in front of the new one and make them run in. Now fill up both hives with frames filled with foundation, and set the new hive on its stand. Perform the operation of dividing in the evening. The way the hives are divided, the new hive has the most of the bees, and the old hive most of the brood ; but the next day the greatest number of the old bees will return to the old hive, and its working force will be greatly increased. The bees in the old hive will proceed to make queen cells, and in about twenty-one days will have a laying queen. If it is desired to utilize the extra queen cells, proceed on the ninth day from the time of making the division, to cut all out but one good one, and give them to queenless colonies or nuclei. Even if you do not want to save them it is best to cut them out, in order to guard against the colony casting a swarm. Cyprian or Syrian colonies in this condition would be most certain to cast a swarm unless the extra cells were removed. If you could give the colony a laying queen it would be much better than allowing it to make one of their own. With bees, *"time is honey."* The objection to placing empty frames in the queenless part of the division would be, that the bees would make drone comb in them till a queen was hatched.

Now suppose you have five colonies in your apiary all near the eve of swarming. Draw two frames (if your new hive will hold 10 frames) of brood and stores with the adhering bees from each, and place them in the empty hive. Replace the frames removed with frames filled with foundation. Be certain not to take any queen. Set the hive on its permanent stand and give the bees a fertile queen. If you can not do this allow them to make queen cells. Here your five colonies contributed to make this new colony, and neither of them sustained any appreciable loss. In a few days time the operation could be repeated. But it is best to bear in mind that the wealth of the apiary must always be measured by the number of strong colonies.

Possibly the highest attainment of an apiarist is skill in breeding a high order of queens. Honey production and breeding queens are to a certain extent antagonistic to each other in practice. The producer of honey must build up his colonies to the greatest strength he is able; while the breeder of queens is continually and unavoidably depleting his colonies and keeping them reduced in strength. Hence the queen breeder is liable to have, in the fall, many weak fragments of colonies that have to be doubled up and fed, at expense, if he wishes to take them over the winter.

Here the question comes up, which is the most profitable, producing honey or breeding queens? This depends upon the extent of pasturage, location and market. If the bee-keeper is in a fair location for honey, he had better sell his honey at ten cents per pound rather than breed queens and sell them at one dollar; for the care, attention, labor and expense attending the production of first-class queens, are infinitely greater.

As the queen is the prime factor in the colony, it is essential that she should possess all the requisites for successfully performing her especial functions. If we desire to improve the qualities of our bees we must commence with the improvement of our queens.

That there is a capacity for improvement in the honey bee I think can hardly be questioned. We know that both plants and animals are endowed with such a capacity, and why should the bee be an exception to this natural law? Our delicious and wholesome apple of the present day was originally the sour, miserable Siberian crab; our sweet and juicy peach was from a bitter fruit of Asia; our improved Irish potato sprang from an insignificant tuber of South America. Our improved breeds of horses, cattle, hogs, poultry, &c., have been brought up to their present state of perfection by intelligently and carefully breeding up the wild originals. How far this improvement can be carried with the bee is difficult to determine, as the organs of reproduction in the queen as well as her fertilization are, I may say, *anomolous*—so unlike the breeding of our domestic animals that the queen breeder will always have immense difficulties to contend with.

In breeding queens of any variety of the honey bee there is a slight tendency to sport and revert back towards the original, especially in the yellow varieties. For this reason it is important to select breeding queens that possess a strong individuality, and capability of stamping their characteristics upon their progeny. This information cannot be obtained by the mere appearance of the queen, but by practically *testing* her queen and worker progeny.

It is very important to select the most desirable drones for the purpose of fertilization. Drones from a vicious and irascible colony may corrupt the worker progeny of the majority of the young queens in an apiary. All impure and objectionable drones should be suppressed by frequently examining the colonies and *shaving* off the heads of the young drones in the cells; by cutting out the drone comb and inserting worker foundation in place, and by the use of drone traps. For this object I recommend Alley's drone and queen trap. Black drones distant four miles from an Italian apiary frequently copulate with the young Italian queens.

The highest type of a queen can only be obtained when all the conditions for her development are the most perfect. These conditions we can learn by observing a hive before the eve of swarming. The hive is crowded with young bees, the temperature is maintained at a uniform heat; honey and pollen are plentiful, and the whole colony is infused with life and and intensity of purpose to perpetuate the race.

Hence it would be but natural for them to put forth their best energies in the development of the future queen that is to lay the eggs to produce the population of the colony. Therefore, the breeder should study the economy and condition of colonies at swarming time, and endeavor to keep his breeding hives in approximate conditions.

Queen cells made in full colonies are generally fine and well formed, and the queen correspondingly fine. This is as they should be under the condition of full colonies. If the cells are examined they will usually be found to be long, rough, with indentations on their surface; and the amount of royal jelly deposited around the embryonic queen to be very abundant. In some cells it is in excess of consumption, and a large quantity is left after the queen crawls out.

Fig. 20

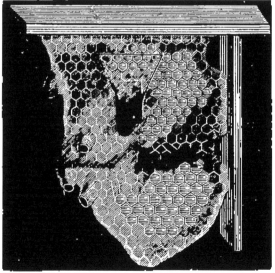

Queen cells inserted in frame of comb, and others in process of development.

Where we have hundreds of queens to furnish, breeding them in full colonies, particularly after swarming time, would be very expensive, hence we must resort to a plan cheaper and more practical.

When a colony is deprived of its queen they instinctively go to work, as soon as the excitement attending the loss subsides, to make another. They are ready to do the work ; but in order to do it well, we must supply them with all the requisites and essentials.

The egg of the queen is analogous to the eggs of fowls and birds. It has its delicate coverings, albumen and yolk; and when the little germ within develops and bursts the shell, it emerges a tiny worm or grub, scarcely discernible with the naked eye. This is now the perfect age of the larvæ for the bees to devevelop into a queen. Always select the larvæ as newly hatched as possible.

At two day's old it makes puny queens, and after the larva gets three days old it is worthless for breeding purposes. It has been demonstrated time and again, that the royal jelly is most abundantly elaborated by young bees, and for this object they must be fully supplied with both honey and pollen. The temperature of the hive must be high enough not to chill the larva.

In order to get the larvæ of the right age you must insert a frame of nice, clean worker comb in the centre of the brood nest of your hive that contains your breeding queen, and if this colony is strong and and the queen prolific—a condition in which it should be kept—the comb may be filled with eggs by the next day; but if the comb is not clean and has been out of the hive for some time, the queen will refuse to lay in it until the bees clean the cells, and, as it were, varnish them. They frequently fill the cells of such comb with honey, rather than have eggs deposited in it.

Observe closely for the eggs; then you must note that these eggs will hatch in three day's time. When hatched, cut out a piece of the comb, say two by four inches, or enough for insertion in the frame on which the queen cells are to be developed. Cut this comb into strips half an inch wide, and cut down the cells on one side to half their depth. The knife for this purpose must be warmed over a lamp so it will easily pass through the comb without clogging or dragging.

Get a frame of tough old comb, free from the eggs and larvæ of the wax-moth, and cut out pieces as long as your strips of brood, but an inch and a half wider, and to the upper edge of these oblong holes you must fasten the strips of larvæ. To do this, have a pan of melted wax and dip the edge of the uncut cells in the wax and immediately apply to the old comb. This will place the larvæ in the cut cells directly downward, with no obstruction beneath, and in the best position for the bees to construct the queen cells. A frame with two cross pieces of wood running from end to end is very convenient for holding the brood. Fasten to these cross pieces an inch strip of old comb, and then to this comb fasten the strips of larvæ as previously directed.

I make my breeding hives only large enough to hold four frames $17\frac{3}{8} \times 9\frac{1}{8}$. At each end I have inch and a half holes for ventilation. These holes are covered with wire cloth and have buttons to close them when necessary. Into one of these hives I hang my frame of selected brood, and also place in two frames, one on each side of the breeding one, filled with honey and pollen but no brood, which I draw from strong colonies. It is very essential that no brood nor eggs should be in these frames; hence observe carefully. The fourth I fill with foundation for the bees to draw out if they will.

To populate this hive with bees of the right age, I go to a strong colony with plenty of young bees, and take out three or four frames with as many young bees as possible and shake all the bees off the combs in front of the breeding hive. Be careful not to get the queen. Stir the bees into the new hive like a swarm. Go to another populous colony and take combs of bees and shake them in front of the hive. Repeat this with other hives, if necessary, until you get into the breeding hive fully two quarts of bees. Close up the entrance with wire cloth and open the ventilators, and carry it, without jarring, into a cool and dark room. Allow it to remain till the evening of the following day, and then, near dark, place it on its stand and open the entrance. Set a board up in front of the entrance to ward off robbers, and to assist the bees in marking the hive. The time this breeding colony was formed must be noted. I use a small slate for registering that is hung on a nail at the front of the hive. On this slate I may write:

        A.
        F.—5th March.
        R.—14th March.

Which is read: Brood from queen, "A"; colony formed March 5th; cells to be removed March 14th, which is nine days after the formation. At this date all the perfect cells, but one, must be cut out and given to queenless colonies. All small, defective cells should be destroyed, for it is better to destroy worthless queens in the larvæ state than to loose time waiting to see if the queens prove good.

Nuclei for the reception of queen cells can be made by drawing frames of brood and stores with all the adhering bees from strong stocks. Select frames with brood ready to crawl out. Two such frames that are well covered with bees placed in a nucleus hive, with a frame of foundation between for the bees to work out, will be sufficient. Such colonies for the reception of the cells should be prepared a day or two before the cells are to be removed; and they should always be kept supplied with brood. The cells should be removed with a small-bladed knife, and with a margin of comb around them. They should be carefully handled, and held in the same position they occupied in the hive.

When the weather is cool, the cells should be inserted in recesses cut in the comb that is in the centre of the cluster, and held in place by thorns until the bees wax them to the comb. When the temperature is warm enough not to chill the brood they can be inserted between the tops of the frames directly over the brood nest. Very frequently the bees will cut the cells when we would suppose from their queenless condition that they would exhibit the greatest care in their preservation. Why they do this is often hard to account for. They are apt to do it when they have laying workers among them.

It is always best to get rid of these laying workers as soon as possible. If the colony has a quart or more of bees, give them a couple of frames of brood from other hives. One frame should have young bees crawling out and the other brood in all stages of development. As soon as the hive gets well stocked with young bees they will most likely, if brood is furnished them, start queen cells. At this stage you can give them a queen cell or a fertile queen. But if the colony has been very much reduced in bees unite them with some other colony.

They are also liable to cut them when not gathering honey and are idle. They are more apt to cut them when they have not a particle of brood in the hive than when they have plenty. I have found by waiting till their own make of cells are capped that they will more readily receive them. But this is too slow for the queen breeder. Time with him is money when he has orders for queens. I have successfully used a wire cage to protect them. This is made out of a piece of stiff wire cloth about $3\frac{1}{2}$ inches square. Make an inch cut in it, one inch from each corner. Bend up the sides to distance of cuts and ravel out two or three of the outside wires. Take one of these wires and secure the corners. Now you have a square wire cup with an open top. Press it firmly into the comb over the cell. The queen cell should be fastened near some cells of uncapped honey so that in case the cell hatches the young queen can get feed. Numerous contrivances have been recommended and used for the preservation of queen cells, but the fuss and bother with their use generally counterbalances all their merit.

In case the cell hatches, release the queen in the same manner as described in directions for introducing queens. I have introduced, and attempted to introduce, thousands of virgin queens by every plan I ever heard of and could think of, and I find there are more failures connected with the introduction of these queens than with fertilized ones. I am aware that there are many bee-keepers who assert that they never have any failures in their introduction; now, all I got to say is, that their experience must be on a very small scale, or their "lucky star" must shine much brighter than mine.

Bees will receive strange queens more readily sometimes than at others. When they are gathering honey plentifully, fertile queens can safely be introduced by most any plan, but when the honey harvest is over, and the bees are comparatively idle, they are often received with very little grace.

Every bee-keeper has his own method of introducing queens which he thinks the best. But it matters not what plan is used, it must be carried out with care and judgment or it will fail. There are many forms and varieties of cages used for this purpose. I prefer one made out of wire cloth with 10 to 12 meshes to the linear inch, and about one and a half inches in diameter with one side open. It has a tin rim soldered around it to hold the wire cloth in position, and resembles in shape a cake-cutter. The cage should be as deep as the comb is thick.

After removing the queen to be replaced, take a frame that has sealed honey from the middle of the cluster, shake all the bees off and press the open side of your cage into the sealed honey at the upper part of the comb. After starting the rim of the cage into the honey, raise up one side of the cage and place the new queen under head foremost. Close down into the comb but in doing so be careful not to catch the queen under the edge of the cage and either maim or kill her. Press the cage into the comb, only leaving a space of half or three-quarters of an inch for the queen. Now replace the frame in the hive and at the end of 48 hours examine to see that all is right. Sometimes the bees will cut the cage loose and kill the queen. To guard against this, confine the cage in position with a pair of transfer sticks. If the bees are clustered on the cage, biting the wires and trying to get at the queen, it would not be safe to release her. Wait till they are evenly distributed on both cage and comb, and shows less anger; then with a small-bladed knife, cut a hole through the comb from the opposite side into the cage, and replace the frame in the hive. Allow the cut and loose particles of comb to remain in the hole to be removed by the bees. In cutting this hole care must be taken not to injure the caged queen. Before cutting the hole be particular to look over the combs and remove every queen cell that they have started. The next day examine the cage again, and if the queen is still in, enlarge the hole a little. Allow the queen to crawl out at will, and do not force her out. The bees that crawl through the hole into the cage will not attempt to hurt the queen, for they seem to be bewildered by the environment and lose sight of her majesty.

It is a not a good plan to use the shipping cage for introducing. This cage becomes filled with bad odors during its passage through the mails, whereby it becomes offensive to the bees, and the confined queen is only received under difficulty. Always use a clean, fresh cage if possible. The queen can be safely transferred to another cage in a close room before a window. Open the shipping cage and let her pass to the window, then catch her by the wing or thorax and place her in the clean cage. Never take hold of her by the abdomen or you might injure her. Do not put any of the bees that come with her in the new cage.

Very valuable queens can be introduced by placing in small nuclei, if the weather is warm enough not to chill the brood. Take a frame of capped brood that is crawling out, and that has unsealed honey, with the adhering bees, carry it a few rods from the hive, give it a gentle shake to get rid of the old bees which will fly back to the hive. Wet the few remaining bees on the frame by a gentle sprinkle, and place the frame in an empty hive with entrances closed with wire-cloth. Drop the queen in and instantly close the hive. Place the hive in a cool dark room. In twenty-four hours add another frame of the same sort of brood and honey, but no old bees. In forty eight hours place the hive on its stand; open entrance to admit one bee at a time, and place some obstructions in front to ward off robbers. The young bees will soon be off to work, and if a frame of brood is added from time to time, it can soon be built up into a strong colony. There are many other plans of introducing queens by spraying, sprinkling, smoking, chloroforming, daubing in honey, &c.; but I deem the above sufficient for all practical purposes.

Mailing cages should be neat, light, ventilated, and provisioned to suit the journey. For distances of 1,000 miles and under, a soft candy made by kneading pulverized sugar into honey, until the latter will take no more, is the best for the trip. In fact, this food if the weather is not too hot, will take queens from the Atlantic coast to the Pacific. But if the weather is very warm, I use a small tin water can with two compartments, in connection with a harder candy made by boiling granulated sugar. With this arrangement, queens will go safely on a twenty or thirty days journey. A cage designed by Mr. Frank Benton is a most excellent one for either short or long journeys. The feed is soft candy.

CHAPTER XI.

DISEASES OF BEES—DYSENTERY—PREVALENT IN THE NORTH BUT UNKNOWN IN THE SOUTH—CAUSE—FOUL-BROOD — APPEARANCE — CAUSE—TREATMENT—INFECTION — TREATMENT OF INFECTED COMBS AND HIVES—BEE PARALYSIS—CAUSES—REMEDIES.

THE honey bee is fortunate in not being subject to many diseases. Dysentery is a disease that often attacks the bees during winter and early spring in northern latitudes, where they are confined for long spells of cold weather; but in the South this disease is not known. It undoubtedly arises from long confinement, with engorgement of the intestines with improper food. The remedies indicated would be: A cleansing flight, warmth in the hive, removal of dampness and better stores.

Foul-brood is not a common disease, but when it takes hold of an apiary it requires watchfulness, great care and determined effort to exterminate it. The cause of the disease is supposed to be a fungus *bacillus alvei* that attacks the larva in the cell before it takes the form of the perfect insect. All the contents of the hive—bees, honey combs, frames and inside surface become infected with the spores which can be conveyed to other hives by inter-passing bees.

The symptoms of the disease, as described by bee-keepers who have contended with it, are: A decline in the prosperity of the colony, and when the brood combs are examined they emit a disagreeable smell from the decomposition of the larvæ. The larvæ turn dark and finally assume a soft, sticky, ropy substance that emits a horrid stench. The capping of the cells of young larvæ is sunk, and often perforated with a little hole.

Mr. Chas. F. Muth, of Cincinnati, and Mr. D. A. Jones, of Canada, have had much experience with this disease. To extirminate the spores Mr. Muth uses salicylic acid, a funcicide, first recommended by a German scientist. His preparation consists of eight grains of salicylic acid, eight grains of soda-borax and one ounce of water. The mixture is applied as follows: First uncap the brood then spray the liquid over the combs with an atomizer. This operation should be repeated three or four times. Mr. Muth found this remedy to work satisfactorily in colonies but slightly affected with the disease, but where the disease is advanced he advises that the bees be transferred into a clean hive filled with foundation, confined, and fed honey or sugar syrup containing a mixture of sixteen grains each of salicylic acid and soda-borax, and one ounce of water to a quart of the syrup.

The treatment of D. A. Jones is more thorough. He removes the bees from the infected hives to a clean empty hive closed with wire cloth. This hive should be placed in a cool dark room. They remain in this hive for 36 or 48 hours till they have exhausted all the honey they had stored in their sacs; then he places them on frames of foundation in a new hive, and feeds. The honey should in no case be fed back to the bees, even if boiled, and this should be done even if used for any other purpose. The combs can be melted up into wax, but the frames had better be buried in the ground or burnt. The hive can be disinfected by washing with carbolic acid, or better with Hydrogen Dioxide, and a coat of paint given inside and out. All these operations must be conducted at such times when no bees can get about to come in contact with the honey, combs, frames or hives, of the infected colony, for if they do, the disease may be communicated to their colony.

Bee-Paralysis is a disease the name of which does not express the symptoms or nature of the disease at all. The disease seems to affect the worker bees in some apiaries at some seasons and then disappears for a time. When first attacked they appear to shake, loose hair, abdomen swells, and they crawl out of the hive as though they were intoxicated and die. All sorts of theories have been advanced as a cause; and just as many remedies have been offered for its cure. But the truth is, we have nothing demonstrative or positive up to the present, either as to cause or remedy. Salt, sulphur, spraying with a disinfectant, changing queens, &c. &c., have had their advocates. After a colony is attacked for a time, the symptoms usually disappear. Some seasons it is more prevalent than at others. From a summing up of the reports of the disease in many different apiaries, I am inclined to offer the opinion (and it is theory) that it is quite likely caused by food infected with microbes, or poisonous mycological formations. It seems to appear more frequently in those apiaries where the bees have access to cider-mills, decaying fruit juice, rotten watermelons, &c. Cheshire claimed that it was caused by *Bacillus Gaytoni*, and yielded to the same remedies as *Bacillus Alvei*.

CHAPTER XII.

ENEMIES OF BEES—THE WAX-MOTH—WHEN INTRODUCED INTO THIS COUNTRY—DESCRIPTION—ITS EGGS AND LARVÆ—GALLERIES AND COCOONS — MOTH-PROOF HIVES—MALLOPHORA—BRAULA COECA OR BEE LICE—ANTS AND TERMITES— HOW TO EXTERMINATE—PROTECTION AGAINST MICE—TOADS DEPREDATORS—SPIDERS—BIRDS.

THE wax-moth, *galleria cereana*, is probably the worst enemy that the Southern bee-keeper has to contend with. This insect is of eastern origin, referred to by Aristotle, and is said to have been introduced into this country in 1805. This miller belongs to the snout family of moths and is of a grey ash color, measuring from head to tip of closed wing about three-quarters of an inch. The wings shut closely on top of the back, slope steeply downwards, and have an upward turn at the end. The female is larger, darker, and has a longer snout than the male.

As soon as evening approaches, this wily insect may often be seen flying around the hives, seeking a place, as near as possible to the waxen cells of the bee, to deposit its eggs. When the hive is densely populated with bees so that all the combs are covered and the entrances well guarded, the moth can not well enter, and its larvæ can seldom effect a lodgment inside of a strong vigorous colony. But if the colony is weak and feeble, and has more combs than the bees can cover, and the approaches to the hive are not well guarded, the moth will soon get inside and lay its eggs on the combs.

When the colony is strong and no entrance can be effected, the moth will lay its eggs in some crack or recess around the hive, usually near the entrance or bottom board. The eggs, when thus deposited, are laid in patches varying from $\frac{1}{8}$ to $\frac{3}{8}$ of an inch in diameter, containing from 75 to 200 eggs. When first hatched the larvæ are scarcely over $\frac{1}{20}$ of an inch long, nearly transparent, and can hardly be seen with the unaided eye. They appear as tiny worms that run with great swiftness. As they can readily pass through the smallest cranny, it is easy to understand how worms often get into those places in the hive that are less frequented by the bees, such as outside cards of comb and sections.

The larvæ are voracious eaters, and grow rapidly. When ready to spin their cocoons they vary in size from $\frac{1}{2}$ to $\frac{3}{4}$ inches in length, depending upon their facilities for perfect development. Their diet is wax. They frequently develop in exposed cakes of rendered wax, but they seem to prefer old comb, and thrive better on it than they do on new. At first their presence can only be told by a little flour-like substance deposited on the comb made by their excrement and particles of wax that they cut. As they proceed with their passage-ways, they line them with silken webs which render them safe from the attacks of the bees. Their heads being covered with a horny shield are proof against stings, whereby they can protude them beyond their silken gallery and proceed with their destructive work with impunity.

The young worms frequently get under nearly mature brood and web the young bees down fast, so that they cannot crawl out, and can only be released by the bees cutting the cells away from around them and pulling out the webs. Such webbed young bees are generally carried out of the hive.

From a knowledge of the natural history of this insect we can easily comprehend the impractibility of all so-call moth proof bee-hives. In fact the only moth-proof hives that can be depended on are those containing strong colonies. In our Southern climate, with our mild winters and long summers, the moth is a much greater pest than it is in colder regions. A greater number of millers always follow a mild winter, and during such years they show themselves in February and March, and keep up a continued series of egg-depositing and transformations until November. Not moth-proof hives, but strong colonies must be the Southern bee-keeper's watchword.

When the colonies are weak, the bees should not be put on more combs than they can cover, and the bottom boards should be kept clean from dirt and the accumulations of wax and pollen.

There are quite a number of insects that prey either upon bees or their stores, but I shall only refer to a few that do injury in our climate. There is a large fly belonging to the genus *Mallophora* that infests some apiaries and destroys many bees. It looks very much like a large bumble-bee, and flies with great swiftness. It takes a position near the hive on some board or limb, and when a bee returns laden with honey it is pounced upon like lightning by this insect, clasped by the legs and carried to some perch where the honey sac of the bee is pierced and its contents sucked out. The bee is then dropped and the fly is ready for another victim.

In some portions of Europe the bees are annoyed by a louse called *Braula Coeca*, but I have never known it to trouble bees in this country. Imported queens from Italy are nearly always infested with these vermin when taken out of the shipping boxes. I have picked off as many as a half-dozen of these lice from a single queen. As I have been exceedingly careful to destroy them before introducing the queen to the colony, I have never found any on any queens or bees reared in my apiary. These lice live on the bee and suck their nourishment from it.

Ants frequently get into the hive and annoy the bees. When they do this they mostly have a nest close by. Hunt it up, open it and scald them with boiling water. Sprinkle around plentifully lime, and the job is completed with a little kerosene painted around the bottom of the hive. Wood lice or termites sometimes cut galleries in the bottom boards and make their nests. Rout them out with kerosene, or by filling their passage ways with pearline.

Roaches at times take up their abode in bee hives to feed upon the honey. A dozen of these insects will consume during the season as much honey as would support a pint of bees. I know of no satisfactory way to get rid of these pests except to frequently open the hives and kill them by hand. Borax has been recommended as a remedy, but I have not found it effectual.

Mice frequently resort to bee-hives during the cold days of winter, and make nests above the bee quilt or honey board, immediately over the cluster, where they can secure the warmth of the bees. When the bees are in a quiet condition the mice cut the combs to get the pollen which they eat; and they will pick off bees from the outside of the cluster and eat the head and thorax and reject the abdomen containing the poison-sac and sting. In this way a single mouse can do great injury. Keep them out by tacking to entrance strips of tin with slots cut in just wide enough to admit a bee but not a mouse. Nail these tins on in the fall before the mice get in.

While toads may be of great benefit to the gardener in the way of destroying noxious insects, they are an injury to the bee keeper. About dark a toad will take its position in front of the entrance and although presenting the picture of innocence, it will take the unsuspecting bee into its capacious mouth with its long tongue with lightning rapidity. It does not seem to care for stings, but gulps the poor bee down as a sweet morsel.

Spiders are considered by some bee-keepers a benefit, as they frequently catch the bee-moth in their webs; but my observation has been that when allowed to spin their webs about the entrance or any part of the hive they catch more bees than moths. Keep your hives clean of spider webs.

Certain birds, as bee-martins, cat birds, sparrows, &c., sometimes catch bees, but as they destroy so many other insects that injure fruit crops, &c., it is best to bear a little with them before we resort to the shot-gun, and endeavor to scare them away without killing them. My apiary is located amid various trees and shrubs, and every season cat birds and mocking-birds come and make their nests and hatch their young in the trees. I have watched them, and at times I thought I saw them catching bees, when I would take my gun and shoot them, and immediately cut open their crop, but I always found the remains of bugs and various other insects, but very rarely a bee and that nearly always a drone. Hence I now never disturb the birds, but let them alone to sing their songs, build their nests, and hatch their young unmolested.

CHAPTER XIII.

BEE-PASTURAGE — DIVERSITY OF MELLIFLUENT PLANTS IN THE SOUTHERN STATES—HOW TO FORM AN ESTIMATE OF THE HONEY-VALUE OF A PLANT—THE PROPER CONDITIONS FOR HONEY SECRETION—SOUTHERN HONEY FLORA—CLASSED AS TO VALUE—HONEY RESOURCES OF FLORIDA—HONEY DEW AND ITS FORMATION.

THE extent and abundance of the honey-producing flora of a country, other conditions being equal, must determine whether apiculture can be successfully and profitably prosecuted in that locality.

Geographically considered, the southern portion of the United States is more varied and diversified in climate, soil and productions than any other. In the mountainous regions of Virginia, North Carolina, Georgia, Tennessee and Alabama, the climate is cool and temperate, and there nearly every plant and fruit that is grown in more northern latitudes can be cultivated to perfection. As we proceed southward, the climate becomes more mild and genial, until we arrive near the Gulf coast, where we approach the "home of the orange." Hence, we perceive that the diversified climate of the Southern States admits of an immense variety of honey-producing plants.

To form a correct estimate of the value of many of our reputed nectariferous plants, would be a very difficult task. In order to arrive at correct conclusions as to the worth of a flower to secrete honey, it requires no little intelligence and accuracy of observation. Most of beginners are too prone to accept for truth the nursery rhyme :

> "How doth the little busy bee
> Improve each shining hour,
> And gather honey all the day
> From every opening flower."

The simple fact of seeing a bee on a flower does not prove that it is gathering one particle of honey. It is bee-nature to hunt for sweets; and in times of scarcity it will visit flowers that it would not touch under more favorable circumstances. Hence, many of the favorable reports of this or that plant for honey are often based upon very hasty and inaccurate conclusions.

To calculate the value of a plant for honey, we must have a sufficient quantity of the same within the immediate range of our bees in order to enable them to work to an advantage The seasons—the atmospheric conditions must not be lost sight of. Too much rain may wash the saccharine secretions away; a protracted drouth may cause its suspension; while a hot, dry atmosphere may evaporate the secretion before the bees can gather it.

When there are many forage plants in bloom at the same time, bees are mostly seen on the ones yielding the most honey; while the rest although secreting some nectar, would be nearly neglected. Therefore, the honey value of some of the trees, shrubs and plants that I shall catalogue as bee forage must necessarily be more or less conjectural.

For the sake of system as well as convenience, I shall divide the honey flora into spring, summer and fall forage. The time and duration of bloom are noted in most cases for the latitude of Augusta, Georgia. North of this point the time will be later, and as we go South the time will be earlier.

The earliest blooming of our spring forage plants is the alder (*alnus*), which commences about the middle of January and lasts, some seasons, till the middle of February. It yields little or no honey, but during bloom its pollen-laden catkins are covered with bees. The amount of pollen that this plant affords is immense, and it comes at a time when breeding should be most encouraged.

In some sections of the South, particularly on light, sandy soils, there may be found some yellow jasmine (*gelseminum sempervirens*). As its flowers possess very decided toxical properties, it is not a very desirable plant to have in range of ones bees. It blooms after the alder, and continues from two to three weeks. Black bees are very seldom seen working on it; but Italians in some seasons, work on it quite briskly. It yields mostly pollen, but very little honey. I have more particularly noted this plant because of its poisonous effect upon young Italian bees immediately after taking their first meal. Since 1871 I have observed commencing with the opening of the yellow jasmine flowers, that there is great mortality among the young bees, which continues until the cessation of bloom, when it ceases as quickly as it came. The symptoms of the poisoning are: their abdomens become very much distended, and they act as intoxicated; there is great loss of muscular power, and they slowly crawl out of the hive and expire. The deaths in strong colonies breeding rapidly in twenty-four hours often amounts to a half pint of bees. What little honey the flowers afford, the bees consume in breeding, and it is very rarely stored, still I know of a few cases of poisoning by eating of gelseminum honey. By the time the bees secure honey from other forage all the jasmine honey, if any, is consumed, and there can be no possible chance for mixture with the main crop.

The wild plum usually commences to bloom the last of February and continues for two or three weeks. Whole acres are often covered with it, forming a dense thicket, and offording the bees rich pasturage. In March we have the peach, wild cherry, and in the latter part of the month the apple, huckleberry, sparkleberry, blackberry, and other plants of minor consideration. Further south they have the ty-ty, saw palmetto and orange, and black mangrove along the coast. The latter affords an abundance of white clear honey of a mild flavor.

The poplar or tulip tree, (*liriodendron tulipifera*), commences to bloom in April, and continues for about three weeks, during which time the bees are kept "booming" carrying in the sweet nectar. This tree is unquestionably the best for forage in the list of southern honey-flora. The honey, while a little dark, is of most excellent flavor.

The holly blooms last for about two weeks and is at its height about the first week in May. The honey is light colored and of good flavor.

In May we have the black gum (nyssa multiflora) and the persimmon. Both excellent for forage. The blooms of both of these trees are *diœcious*, that is, the male flower is found on one plant and the female flower on another. The bees work more on the male than they do on the female flowers.

The bay (magnolia glauca) in some seasons yields large quantities of honey of a good quality. This tree flowers for at least one month and extends into June. The magnolia grandiflora, linden and honey-locust, also bloom in May. The latter I regard as a most valuable forage tree. During its time of bloom bees swarm on it the entire day to the neglect of other forage.

The china tree (melia azederach) affords some honey. Its period of bloom is about two weeks.

Sourwood, varnish tree (sterculia platanifolia), Japan privit (lugustrum), and a few other plants of minor consideration constitute the principle forage in June. I have now enumerated the chief honey-producing plants that go to make up our spring harvest. Take one season with another, bees commence to lay up supplies about the middle of April, and continue till middle of June. After this date but little honey is gathered till fall.

There is comparatively little forage during the summer months of July and August. The button bush (cephalanthus occidentalis), sumac and asclepias tuberosa (sometimes called butter fly weed), are the most important. Sumac yields honey abundantly, but a warm, dry atmosphere evaporates it very rapidly so that bees only work on it very early in the morning.

A noxious weed, know as *hellenium tenniefolium*, that has made its appearance in the Southern states since the Confederate war, yields a very bitter, yellow looking honey. It blooms in July and August, has a yellow flower, grows along roadsides and in uncultivated places. It is often taken for a species of dog-fennel, but it is altogether distinct. The honey from this plant will spoil the flavor of the spring crop if any of the latter is left in the hive. For breeding and wintering it answers all the purposes of a better article, but it is worthless for market. Bees are very rarely seen on this weed after the appearance of fall pasturage.

In some seasons the cotton bloom yields honey, which is of a light amber color and of good flavor. But generally they gather more pollen than nectar from the flowers of this plant.

Bees work with considerable energy on the cow-pea, and gather some honey, though I do not think this plant yields large quantities. There is a peculiarity in the honey-secreting organs of this plant, in the fact that they are contained in little glands located on the peduncle or flower-stem and not within the flower.

Goldn rod (solidago) and the asters bloom in September and continue till frost. In some sections of country the golden rod is esteemed a valuable forage plant, but my observations do not confirm this idea. The aster, while a modest and unpretentious little flower, is the most valuable fall forage bloom we have. The honey is of a very light amber color and very fine flavor. In some seasons the bees store large quantities of surplus from this source alone.

CULTIVATED FORAGE.

I am satisfied that it will never pay to cultivate plants exclusively for the honey. To be profitable, they must have other uses besides the honey. Both red and white clover do well on our clay and sandy loam soils that are sufficiently rich for their growth, but it is folly to expect success on light sandy uplands. Alsike and melilot have been successfully grown in a few favored localities where the conditions for growth have been favorable. It is not only difficult to get a stand, but even after a possible stand is obtained, the plants are killed during our long, dry summer.

Crimson clover and alfalfa have been tried on both clay and sandy loam soil and have proved a success.

Buckwheat grows well, but it either fails to secrete honey during the summer months or the honey is dissipated by the dry, hot air before the bees can gather it. If sown so as to bloom in either spring or fall, it comes in competition with plants that are richer in nectar, and the bees refuse to work on it.

Cat-nip, horse-mint, mustard, rape and turnip blooms, when cultivated, yield much honey. The first two of these plants could profitably be cultivated in all out-of-the-way places.

When laying out pleasure grounds and planting shade trees, it would be advisable to keep an eye to utility as well as to ornament. Many of the most valuable and ornamental shade trees are also excellent for bee forage. I can especially recommend the paulonia, catalpa, chinaberry tree, varnish tree and mimosa. I think the most of these beautiful trees are natives of Japan, a country to which America is greatly indebted for a large number of her most highly ornamental trees and plants.

When we get into the semi-tropical region of Florida, the forage, as well as its time of bloom, differs from that of higher latitudes. Mr. W. S. Hart, a most successful bee-keeper of Hawk's Park, Florida, writes: "The first honey flow of the season comes from soft maple in January, followed by yellow jasmine, willow and orange blossoms in February and

March, which is used for brood raising. April is always dry, and but little honey is gathered. About the second week in May the gall berry and saw palmetto come in, and give a surplus of fine amber honey of heavy body and good flavor. Early in June the black mangrove (*avicennia tomentosa*) comes in, and everything else is neglected by the bees, until the cabbage palmetto (*sabal palmetto*) comes in July when they gather from both sources. The honey however, is so nearly identical that only an expert can tell them apart. The last, having a trifle more color. I have never seen a handsomer or clearer honey than that from the black mangrove, and the flavor is mild and very fine. It is a honey that every one likes, and will wear without cloying on the taste better than almost any honey produced in America. Its body is not as heavy as that of California sage or white clover, yet it is not lacking in that respect if taken after being capped, or cured after extracting. It lasts until 1st to 10th of August. The cabbage palmetto is about the same date. Then comes a honey drouth until the middle of September, when wild sunflower, and many other plants give a flow that produces some surplus throughout the fall and early winter."

There are some species of aphides or plant lice that infest some varieties of trees and shrubs, that eject a saccharine excrementitious matter in the form of a fine spray that collects on the leaves and plants beneath them. This is called *honey dew*, and when the bees have no other forage, they collect large quantities of this honey, but it is a very inferior quality.

CHAPTER XIV.

MARKETING HONEY—THE PEOPLE MUST BE EDUCATED TO A FULL APPRECIATION OF THE USES OF HONEY—STRAINED AND EXTRACTED HONEY—GRANULATION NO SIGN OF IMPURITY—HOW TO PREPARE IT FOR MARKET—HOW TO OFFER IT—TO WHOM TO SHIP—GLUTTING THE MARKET.

IN some sections of country it is much easier to produce the honey than it is to find a market for it. Where such is the case, the producers must bring to bear more energy and endeavor to create a market for it. There are few places where honey can not be sold if proper means are used to call attention to its merits. The low dark grades of it are now largely used in the arts—by brewers, tobacconists, bakers, &c. Educate the tastes of the people to an appreciation of it by distributing among them tracts, explaining its medicinal and culinary uses, and its great wholesomeness and superiority over the bulk of the common syrups on the market. The most of these syrups are vile adulterations of cane syrup with glucose, too unwholesome to be taken into the human stomach.

The wide-awake bee-keeper will study the demands of his market, and then secure his honey in the neatest condition to supply those demands. Some customers may want extracted honey, while others will have none but comb honey. It will be necessary to explain the difference between extracted honey and strained honey. One is honey slung out of the comb with a machine by centrifugal force, leaving behind the comb, pollen and brood, while the other is strained through a cloth or sieve from mashed combs containing pollen, larvæ, dirt, &c. I regard extracted honey more wholesome than comb honey, from the fact that the wax is indigestible in the human stomach, and would more than likely produce irritation. This is why some persons say that honey does not agree with them—they eat the wax.

I find for a near market that it is best to put up extracted honey in pint, quart and half gallon vessels. Muth's jars are excellent for small packages; so are also self-sealing fruit jars and small tin pails or buckets. Have a neat label with your name on each package.

It must be remembered that nearly all honey will gradually granulate in cold weather. Granulation is no sign of adulteration; but on the contrary in some parts of Europe it is taken as a sign of purity. Granulated honey can be reduced to liquid by placing the vessels containing it in boiling water. If heated above a boil, it will lose its flavor and deteriorate in quality.

Extracted honey should be shipped to distant markets in cypress kegs and barrels. These are much better than those made out of hard wood. A very thin coating of wax or paraffine may be given to the inside of the barrel, but a thick coat is worse than no wax at all, from the fact that it will crack and break loose. A coat of shellac varnish is preferable to wax, and is an effectual preventive of leaks; but no inside coating can be properly applied unless the vessel is perfectly dry on the inside. After the application of the coating, the hoops must be driven up tight.

Comb-honey sells best in small frames or sections holding about one or two pounds. When in these small neat packages, it is more inviting than when offered for sale in bulk in bucket, tubs or barrels, as is frequently done by the old style bee-keepers. A person might be tempted to invest ten or twenty cents for a pound or two in a neat section free from daub, when he would refuse to invest the same amount for honey in an uninviting condition.

Encourage and build up your home markets. Better sell for a little less near home than ship to distant markets and pay freights and commissions. Bee-keepers frequently consign their honey to ordinary grocers and commission merchants who have not the remotest idea how to handle honey to the best advantage. It frequently arrives in bad condition, and then it goes to the cellar or back part of the store, out of the way, where no one can see it or know that they have it. If you want your grocer to handle it profitably to yourself, provide him a neat tight glass case in which to place it and exhibit it to his customers.

When shipping to distant markets, first find out the responsibility of the party to whom you ship, and his capability and facilities for handling the article. If you have no way to get at this information, correspond with producers who effect their sales through commissions houses.

In places where there is only an ordinary demand for honey, caution should be used not to glut the market. Only offer as much at a time as you can readily sell. Offer frequently, and your sales will increase at a paying price, and during the year a very large quantity can be disposed of.

CHAPTER XV.

USES OF HONEY IN MEDICINAL PREPARATIONS—IN COOKING AND IN ARTS—REMEDIES FOR DISEASES OF THE MOUTH, THROAT, BRONCHI AND LUNGS —LAGRIPPE AND COLDS—RECEIPTS FOR HONEY CAKES, GINGER SNAPS, COOKIES, PUDDINGS, VINEGAR, METHEGLIN, MEAD, &c.

IN the preparation of many medical compounds honey plays a very conspicuous part, and in the culinary art it can be made to occupy a most important place. In the *materia medica* it is classed among the vegetable cathartics. It enters into many pulmonary and cough mixtures for diseases of the throat, bronchi and lungs. When combined with a little flour, it is one of the best applications for boils, wounds, scalds and burns. It also enters into the composition of some of the tooth pastes or dentrifices.

In order to encourage a more general use of honey, I append a few receipts or formulas for the invalid and for the kitchen wherein honey forms a prominent part. For the most of them I am indebted to the bee journals, and some of them I have picked up as waifs floating upon the stream of periodical literature, while a few I have formulated and proved good under my own "vine and fig tree."

HONEY MOUTH WASH.

This mixture is excellent in aphthous sore mouth of children, and also for cracked tongue: One ounce of sage leaves to a pint of boiling water; 3 scruples of borax and 2 ounces of honey.

FOR APTHOUS AFFECTIONS OF THE MOUTH.

One drachm of borax rubbed up with one ounce of honey.

For relaxation of the uvula and as an astringent wash in mercurial sore mouth: ½ ounce alum to 1 pint of water sweetened with honey.

HONEY CANDY,

For la grippe and colds attended with sore throat: 1 pound granulated sugar, ½ pound honey, two tablespoonsful decoction of horehound; add enough water to wet the sugar; boil without burning until it will candy, then add teaspoonful of juniper tar.

HONEY COUGH MEDICINE.

It is especially recommended for long standing coughs: Extracted honey, linseed oil, whiskey of each, 1 pint; mix. Dose, one tablespoonful 3 or 4 times a day.

CROUP REMEDY.

This is good in all cases of mucus and spasmodic croup:

Raw Linseed Oil	2 ozs.
Tincture of Blood Root	2 drs.
Tincture of Lobelia	2 drs.
Tincture of Aconite	½ dr.
Honey	4 ozs.

RESIN CERATE.

For burns, wounds, &c.: Rosin, five ounces; lard, eight ounces; yellow bees-wax, two ounces. Melt together and stir constantly until cold. This preparation is most highly recommended.

HONEY CAKE.

One pint flour, one tablespoonful of butter, one teaspoonful of soda, two teaspoonsfuls of cream of tartar, and honey sufficient to make a thick batter; spread about an inch thick and bake in a hot oven.

HONEY SPONGE CAKE.

One large coffee cup full of honey, one cup of flour, 5 eggs. Beat yolks and honey together, beat the whites to a froth; mix all together, stiring as little as possible; flavor with lemon juice or extract.

HONEY CAKE.

One quart of extracted honey, one-half pint sugar, $\frac{1}{2}$ pint melted butter, 1 teaspoonful soda dissolved into $\frac{1}{2}$ tea cup of warm water, $\frac{1}{2}$ of a nutmeg, and 1 teaspoon of ginger. Mix these ingredients and then work in flour and roll. Cut in thin cakes and bake on buttered tins in quick oven.

RAILROAD HONEY CAKE.

One cup of honey, 1 heaping cup of flour, 1 teaspoonful cream tartar, $\frac{1}{2}$ teaspoonful soda, 3 eggs and a little lemon juice; stir together 10 minutes. Bake in a quick oven.

GERMAN HONEY CAKE.

Three and one-half pounds of flour, $1\frac{1}{2}$ pounds of honey, $\frac{1}{2}$ pound sugar, $\frac{1}{2}$ pound of butter, $\frac{1}{2}$ of a grated nutmeg, $\frac{1}{6}$ ounce of ginger, $\frac{1}{4}$ ounce of soda; roll thin, cut in small cakes and bake in a hot oven.

HONEY FRUIT CAKE.

Four eggs, 5 cups of flour, 2 cups of honey, 1 teacupful of butter, 1 cup of sweet milk, 2 teaspoonfuls of cream of tarta, 1 teaspoonful of soda, 1 pound of raisins, 1 pound of currants, $\frac{1}{2}$ pound of citron, 1 teaspoonful each of cloves, cinnamon and nutmeg; bake in a large loaf in a slow oven. This will be nice months after baking as well as when fresh.

HONEY LEMON CAKE.

One cup butter, 2 cups honey, 4 eggs well beaten, teaspoonful essence of lemon, half a cup of sour milk, 1 teaspoonful of soda, flour enough to make it as stiff as can be stirred; bake at once in a quick oven.

HONEY APPLE CAKE.

Soak three cups of dried apples over night; chop slightly, and simmer in 2 coffee cups of honey, ½ coffee cup of sugar, 1 coffee cup of melted butter, 3 eggs, 2 teaspoonsful of soda, cloves, cinnamon, powdered lemon or orange peel, and ginger syrup if you have it. Mix all together, add the apples and then flour enough for a stiff batter; bake in a slow oven. This will make two good sized cakes.

CHEAP HONEY CAKE.

One teacup of extracted honey, ½ teacup of thick sour cream, 2 eggs, ½ teacup of butter, 2 cups flour, scant ½ teaspoon of soda, 1 teaspoon of cream tartar; flavor to taste.

HONEY GINGER CAKE.

Three cups of flour, 1½ cups of butter; rub well together, then add one cup brown sugar, 2 large tablespoonsful of ginger, 5 eggs, 2 cups of extracted honey and 3 teaspoonsful of baking powder. Beat it well, and bake in a square iron pan one hour or more.

HONEY TEA CAKES.

Three pounds and a half of flour; 1½ pounds of honey; ½ pound of sugar; ½ pound of butter; ½ a nutmeg grated; 1 tablespoonful of soda dissolved in a little hot water. Roll it a quarter of an inch thick; cut it into small cakes, and bake them 25 minutes in a moderate oven.

HONEY COOKIES.

Mix a quart of extracted honey with ½ a pound of powdered white sugar, ½ a pound of fresh butter, and the juice of two oranges or lemons. Warm these ingredients slightly, just enough to soften the butter, and then stir the mixture very hard, adding a grated nutmeg. Mix in gradually a pound or less of sifted flour, make it into a dough, just stiff enough to roll out easy, and beat it well all over with a rolling pin; then roll it out into a large sheet half an inch thick, cut it into round cakes with the top of a tumbler, dipped frequently in flour, lay them in shallow tin pans slightly buttered and bake them.

HONEY CAKE

Three cups of honey, 4 cups sour milk, ½ cup butter, soda to sweeten the milk; mix rather stiff.

HONEY GINGER SNAPS.

One pint honey, ¾ pounds of butter, 2 teaspoonfulls of ginger, boil together a few minutes, and when nearly cold put in flour until it is stiff, roll out thinly and bake quickly.

HONEY PUDDING.

Three pints thinly sliced apples, 1 pint honey, 1 pint flour, 1 pint corn meal, small piece butter, 1 teaspoonful soda, the juice of two lemons and thin grated rinds; stir the dry soda into the honey, then add the apples, melted butter and a little salt; now add the lemon rind and juice and at once stir in the flour. Bake one hour. Serve hot or cold with sauce.

VINEGAR.

Stir together a half pound of honey and a quart of water, permitting the whole to boil while mixing it; then expose it to the rays of the sun; covering with light muslin to prevent insects from getting in, and in 6 weeks it will become excellent vinegar, quite as good flavored as that made from wine.

METHEGLIN.

Honey, fourteen pounds; warm water, three gallons; yeast, half gill; two ounces hops boiled in a quart of water. Mix the water, after straining out the hops, with the rest of the material. Put all into a cask or demijohn, and add enough water to make the whole four gallons; let it work three days, then bottle and tie corks.

MEAD.

Twelve gallons water; whites of six eggs; mix well; then add twenty pounds of honey; boil one hour; then add cinnamon, ginger, cloves, mace, rosemary; as soon as cold, put one spoonful of yeast in it. Barrel, keeping the vessel full as it works. After working stop close. When fine, bottle for use.

CHAPTER XVI.

APIARY WORK PLANNED FOR THE YEAR.

IN mapping out the work in the apiary for the year, it should be borne in mind that it is done in reference to the latitude of Augusta, Ga.

In this latitude bees require no special hives or repositories to enable them to pass the winter in safety. Double wall and chaff hives are not necessary. Neither is it necessary to place them in cellars, caves, clamps or pits, as is required in the North, in order to carry them safely over the cold winter. Here they winter in perfect security on their summer stands without any protection whatever. As they can fly out every few days during our winters and void their faeces, they never suffer from attacks of dysentery, a disease that often prevails among the bees during wintering in the North. But notwithstanding our mild winters, it is essential, in order to secure the best results, for our bees to go into winter quarters in proper condition if we wish them to come out strong in the spring. That they should have sufficient stores to supply the demands of the colony, and to carry them until the middle of March or first of April is a *sine qua non*. Every good, strong colony will need from 20 to 30 pounds of honey to see them safely over the winter and bad days of spring. A weak colony will consume more honey in proportion than a strong one, hence it is best to have in the fall as few of the former as possible.

In October examine all your colonies and if they

have not sufficient stores for the winter feed up at once. It is now too late to build up weak stocks, hence such had better be united. Remove all sections and cover the brood frames with a honey-board having a bee space beneath, or cover with a quilt made out of a piece of coarse cloth. Burlaps answer the purpose most admirably, and I much prefer quilts of this material to enamel cloth which cost much more.

Under the cloth and across the tops of the frames, place a couple of short sticks to afford a passage-way for the bees to pass over the top bars from one comb to another without crawling around at an expense of animal heat. Tuck the edges of the quilt down tightly so no bees can pass above. Tack slotted tins in front of the entrance to keep out mice. When thus prepared, do not disturb them any more until February, when it is necessary to examine the condition of every hive in order to see what progress has been made at breeding, and to clean off all deposits of wax and debris on the bottom board.

Bees will winter safely on their summer stands without extra protection anywhere south of the latitude of Maryland. Of course it is always best to have them sheltered from the cold winds by a fence hedge or break. Farther north they should have some extra protection. Double walled hives with space packed with chaff, or some nonconducting substance, are now pretty generally conceded by northern bee-keepers to be the best for their winters. The bees are wintered in these hives on their summer stands. On the other hand, cellar wintering is practiced just as successfully by others. Successful wintering does not depend so much upon in-doors or out-doors, as it does upon an observance of the conditions upon which the health and integrity of the bees depend. First, the colony must have from twenty to thirty pounds of honey well cured and of good quality. In the South the quality of the stores is not as important as it is in cold climates. The hive or repository should be free from damp, and maintained at a temparature of about 45°. The place should be kept dark, and the bees free from all jars and sudden motions.

Mr. W. Z. Hutchinson, in his work on "Advanced Bee-Culture", remarks on wintering bees in the North, that. "unless the cellar is well under ground, where it is well beyond the influence of the outside temperature, it is well to keep watch and not allow the temperature to run too low in protracted cold spells. A lamp stove burned all night in a cellar will raise the temperature several degrees. During the fore part of winter a low temperature is not so dangerous as it is toward spring, when brood rearing has commenced. From 35° to 45° will do very well until towards spring, when it should not be allowed to go below 40°, and may with safety go as high as 48° or 50°. In this connection it must be remembered that moisture has an influence upon the effects of temperature. So far as effects are concerned a moist atmosphere is the equal of a low temperature.

If the cellar is moist, either raise the temperature or remove the moisture. Unslacked lime in the cellar will absorb moisture. Even when the influence of moisture has been considered, it will not answer to tie ourselves to a certain temperature. It is the temperature *inside* the hives that affect the welfare of the bees. If the colonies are weak, their hives open and the brood nest uncontracted, a higher degree of heat is needed than with strong colonies in close, well protected hives. Putting colonies near the top of the cellar will help matters some as the air is warmer there. The best guide in regard to this matter of temperature is the behaviour of the bees themselves. If they are closely, quietly and compactly clustered, there is but little cause for alarm in regard to the temperature. Quite a number have reported excellent results by warming up the bee-repository to summer heat, say once a week or ten days; if the bees become uneasy toward spring. This enables the bees to throw off any surplus moisture, and, as the temperature goes down, they quiet down and remain so for several days, when they may be warmed up again. So long as the bees remain quiet, I should not disturb them by artificial heat. If the cellar becomes *too warm* in the spring, before it is time to remove the bees, it may be cooled down by carrying in snow or ice, or the window or doors may be opened at night and closed in the morning."

"Years ago many bee-keepers practiced taking their bees from the cellar if there came a warm day in the winter, and allowing them to fly, returning them again to the cellar, but this practice has been pretty nearly abandoned. If the bees are in a quiet, normal condition it often arouses them and sets them to breeding in mid-winter, which is far from desirable. If the food, temperature and other surroundings are what they ought to be, such a flight is not needed. If they are very faulty, such a flight will not save the bees from death."

"If bees out of doors are properly protected and have abundant stores, they need no care in winter, unless it is to see that the entrances are not clogged with ice, snow or dead bees when there comes a day warm enough for them to fly. If a rim two inches wide is put under each hive when they are packed in the fall, and an entrance made at the *upper edge* of this rim, the entrance will never be clogged with dead bees."

"In my opinion," continues this writer, "food is the pivotal point upon which turns the wintering of bees in our Northern States. Food is the fulcrum, and temperature the long end of the lever."

"The whole question in a nut shell is just this: The loss of bees in winter, aside from that caused by diarrhœa is not worth counting. It is *diarrhœa* that kills our bees. What causes it? An overloading of the intestines with no opportunity for unloading them. Cold confines the bees to their hives. The greater the cold the larger are the quantities of food consumed to keep up the animal heat. The more food there is consumed, the sooner are the intestines overloaded. Doesn't it seem clear that the character of the food consumed would have an effect upon the amount of accumulation in the intestines? In the digestion of cane sugar there is scarcely any residue. Honey is generally quite free from nitrogenous matter, being well supplied with oxygen, and when free from floating grains of pollen is a very good and safe winter food, although not as good as properly prepared sugar syrup, which never contains nitrogen but does possess more oxygen. The excrete from diarrhetic bees is almost wholly pollen grains, in a digested or partly digested state, with a slight mixture of organic matter. What overloads the intestines of the bees is this nitrogenous matter that they consume, either as grains of pollen floating in the honey, or by eating the bee-bread itself."

Winter is the best time to order hives and material

for the apiary. At this season supplies can generally be bought cheaper; and it gives time for delays in shipping, and time for the bee-keeper to properly set up his hives and paint them, so as to be ready when he wants to use them. It is a great mistake to wait till swarming time or till the honey harvest has commenced before you order your supplies.

The long winter evenings offer an excellent time to read up on bee-culture—to read the periodicals and books pertaining to the subject.